Lecture Notes in Mathematics

Volume 2361

Editors-in-Chief

Jean-Michel Morel, City University of Hong Kong, Kowloon Tong, China
Bernard Teissier, IMJ-PRG, Paris, France

Series Editors

Karin Baur, University of Leeds, Leeds, UK
Michel Brion, UGA, Grenoble, France
Rupert Frank, LMU, Munich, Germany
Annette Huber, Albert Ludwig University, Freiburg, Germany
Davar Khoshnevisan, The University of Utah, Salt Lake City, USA
Ioannis Kontoyiannis, University of Cambridge, Cambridge, UK
Angela Kunoth ⓘ, University of Cologne, Cologne, Germany
Ariane Mézard, IMJ-PRG, Paris, France
Mark Podolskij, University of Luxembourg, Esch-sur-Alzette, Luxembourg
Mark Policott, Mathematics Institute, University of Warwick, Coventry, UK
László Székelyhidi ⓘ, MPI for Mathematics in the Sciences, Leipzig, Germany
Gabriele Vezzosi, UniFI, Florence, Italy
Anna Wienhard, MPI for Mathematics in the Sciences, Leipzig, Germany

This series reports on new developments in all areas of mathematics and their applications - quickly, informally and at a high level. Mathematical texts analysing new developments in modelling and numerical simulation are welcome. The type of material considered for publication includes:

1. Research monographs
2. Lectures on a new field or presentations of a new angle in a classical field
3. Summer schools and intensive courses on topics of current research.

Texts which are out of print but still in demand may also be considered if they fall within these categories. The timeliness of a manuscript is sometimes more important than its form, which may be preliminary or tentative. Please visit the LNM Editorial Policy (https://drive.google.com/file/d/1MOg4TbwOSokRnFJ3ZR3ciEeKs9hOnNX_/view?usp=sharing)

Titles from this series are indexed by Scopus, Web of Science, Mathematical Reviews, and zbMATH.

Augustin Banyaga • David Hurtubise • Peter Spaeth

Twisted Morse Complexes

Morse Homology and Cohomology
with Local Coefficients

 Springer

Augustin Banyaga
Department of Mathematics
Penn State University
University Park, PA, USA

David Hurtubise
Department of Mathematics and Statistics
Penn State Altoona
Altoona, PA, USA

Peter Spaeth
Nondestructive Evaluation Sciences Branch
NASA Langley Research Center
Hampton, VA, USA

ISSN 0075-8434 ISSN 1617-9692 (electronic)
Lecture Notes in Mathematics
ISBN 978-3-031-71615-7 ISBN 978-3-031-71616-4 (eBook)
https://doi.org/10.1007/978-3-031-71616-4

Mathematics Subject Classification: 37D15, 55N25, 57R19, 55P45, 57R70, 58E05, 55N30

© The Editor(s) (if applicable) and The Author(s), under exclusive license to Springer Nature Switzerland AG 2024

All rights are solely and exclusively licensed by the Publisher, whether the whole or part of the material is concerned, specifically the rights of translation, reprinting, reuse of illustrations, recitation, broadcasting, reproduction on microfilms or in any other physical way, and transmission or information storage and retrieval, electronic adaptation, computer software, or by similar or dissimilar methodology now known or hereafter developed.
The use of general descriptive names, registered names, trademarks, service marks, etc. in this publication does not imply, even in the absence of a specific statement, that such names are exempt from the relevant protective laws and regulations and therefore free for general use.
The publisher, the authors and the editors are safe to assume that the advice and information in this book are believed to be true and accurate at the date of publication. Neither the publisher nor the authors or the editors give a warranty, expressed or implied, with respect to the material contained herein or for any errors or omissions that may have been made. The publisher remains neutral with regard to jurisdictional claims in published maps and institutional affiliations.

This Springer imprint is published by the registered company Springer Nature Switzerland AG
The registered company address is: Gewerbestrasse 11, 6330 Cham, Switzerland

If disposing of this product, please recycle the paper.

Preface

This book gives a detailed presentation of Morse homology and Morse cohomology with local coefficients—the so-called "twisted Morse homology and cohomology."

Morse theory connects algebraic and differential topology. A fundamental fact is that every finite dimensional closed smooth manifold M has the homotopy type of a CW-complex, with k-cells in one-to-one correspondence with critical points of index k of a Morse function on M. Furthermore, the singular homology of M with integer coefficients is isomorphic to the homology of the Morse-Smale-Witten chain complex, generated by the critical points of a Morse function on M, i.e., the "Morse homology" [8].

Steenrod credits Reidemeister for introducing local coefficients into algebraic topology in 1938 [70, 84]. However, it was Steenrod who published the first extensive survey of singular homology and cohomology with local coefficients in 1943 [83]. Subsequently, Steenrod defined cellular chain and cochain complexes with local coefficients for topological spaces with a cellular decomposition via a cellular boundary operator analogous to the CW-boundary operator. However, a cellular decomposition is composed of cells that are homeomorphic to closed simplices and is hence less general than a CW-structure, cf. section 31 of [84].

Singular homology with local coefficients in an orientation bundle was used by Bott in the context of Morse-Bott theory to "count" critical submanifolds when the unstable normal bundles of the critical submanifolds are not orientable [16]. Subsequently, various applications of local coefficient systems in the context of Morse theory and Floer theory have been studied.

Considering this history, extending the Morse-Smale-Witten chain complex to allow for general local coefficient systems presents itself as a rather obvious project. However, there are some technical issues that arise when carrying out this project using classical techniques from algebraic topology and homotopy theory. Most of these issues arise from the fact that a Morse function determines a CW-complex, but Steenrod's cellular boundary operator with local coefficients is not well-defined for general CW-complexes. However, it is defined for *regular* CW-complexes.

The cellular boundary operator with local coefficients is well-defined for a regular CW-complex because there is a unique homotopy class of paths between the basepoints of adjacent cells of relative dimension one. On the other hand, the Morse-Smale-Witten boundary operator with local coefficients is defined for any Morse-Smale pair (f, g) on M, regardless of whether or not the CW-complex determined by the pair is regular, because the gradient flow lines serve as preferred paths for the local system.

In this book, we prove that the homology of a Morse-Smale-Witten chain complex with local coefficients is isomorphic to the singular homology of M with local coefficients by first showing that the homology of the Morse-Smale-Witten chain complex with local coefficients does not depend on the Morse-Smale pair used to define the complex. We then construct a Morse-Smale pair whose unstable manifolds determine a regular CW-structure on M. In fact, we prove that there is a Morse-Smale pair (f, g) whose unstable manifolds coincide with a smooth triangulation of M, and the Morse-Smale-Witten boundary operator with local coefficients for (f, g) coincides with Steenrod's cellular boundary operator with local coefficients.

Twisted Morse complexes are effective tools for computing homology and cohomology with local coefficients, because they often have fewer generators than chain complexes coming from cellular decompositions.

To demonstrate this, the Lichnerowicz cohomology of a surface of genus two is computed explicitly with respect to all closed 1-forms on the surface, and following a result due to Albers et al. [2], twisted Morse complexes are used to show that certain manifolds are not associative H-spaces. We also discuss the Novikov homology of a closed 1-form and how it relates to twisted Morse homology. We then use twisted Morse complexes to compute the Novikov numbers of various surfaces with respect to all closed 1-forms on the surfaces.

Acknowledgments

We would like to thank the following people for several useful conversations on various aspects of this project: Pinaki Das, Wojciech Dorabiala, Mark Johnson, Thomas Krainer, and Karl Lorensen. In particular, the analytic approach to the invariance of the Euler number for Lichnerowicz cohomology (Remark 5.26) was brought to our attention by Thomas Krainer, and the simple yet elegant argument proving that rank one cohomology classes are dense (Remark 6.24) comes from Pinaki Das. We would also like to thank the anonymous referees for their many useful comments and suggestions, which significantly improved this book. In particular, we would like to thank the anonymous reviewer who suggested the alternate approach to proving Theorem 4.1 outlined in Remark 2.23.

University Park, PA, USA Augustin Banyaga
Altoona, PA, USA David Hurtubise
Hampton, VA, USA Peter Spaeth
July 2024

Contents

1 Introduction 1
   1.1 Morse Homology with Integer Coefficients 1
   1.2 Extensions and Applications 3
   1.3 Summary of Main Results 4

2 The Morse Complex with Local Coefficients 11
   2.1 Local Coefficients 11
   2.2 Path Components of Compactified Moduli Spaces 13
   2.3 Twisting the Morse-Smale-Witten Boundary Operator 16
   2.4 Examples 18
   2.5 Computation of $H_0((C_*(f;G), \partial_*^G))$ 23
   2.6 The Morse Eilenberg Theorem 24
       2.6.1 The Morse-Smale-Witten Chain Complex on a
             Covering Space 25
       2.6.2 The Morse Eilenberg Theorem 29

3 The Homology Determined by the Isomorphism Class of G 37
   3.1 A Chain Map 37
   3.2 A Chain Homotopy 42
   3.3 An Invariance Theorem 49

4 Singular and CW-Homology with Local Coefficients 51
   4.1 Singular Homology with Local Coefficients 52
   4.2 Regular CW-Complexes 53
   4.3 Unstable Manifolds and Regular CW-Structures 58
   4.4 A Morse-Smale Function that Determines a Regular
       CW-Structure 61
       4.4.1 Outline of the Proof 62
       4.4.2 The Construction on a 2-Simplex 64
       4.4.3 The Construction for Adjacent 2-Cells 68
       4.4.4 Proof of Theorem 4.12 71
   4.5 Local Coefficient Systems of R-Modules and the Euler Number 83

5 Twisted Morse Cohomology and Lichnerowicz Cohomology 89
- 5.1 Twisted Morse Cohomology 89
- 5.2 Lichnerowicz Cohomology and LCS Manifolds 92
- 5.3 Mapping Differential Forms to Morse-Smale-Witten Cochains 96
- 5.4 Relationship to Sheaf Cohomology 104

6 Applications and Computations 107
- 6.1 Parallel 1-Forms and Lichnerowicz Cohomology Computations 107
- 6.2 H-Spaces... 118
- 6.3 Novikov Homology 124
  - 6.3.1 A Covering Space Associated to a 1-Form 125
  - 6.3.2 Novikov Rings 128
  - 6.3.3 A Local Coefficient System of Rank One Nov-Modules 128
  - 6.3.4 Novikov Homology... 130
  - 6.3.5 Novikov Numbers 135
  - 6.3.6 Novikov Inequalities.. 136
  - 6.3.7 Novikov Numbers and Twisted Morse Homology 137

References 147

Index 153

Chapter 1
Introduction

1.1 Morse Homology with Integer Coefficients

Morse homology is a homology theory defined using a smooth Morse-Smale function $f : M \to \mathbb{R}$ on a smooth Riemannian manifold $(M, \mathfrak{g})$. The k-th chain group $C_k(f)$ is defined using the critical points of index k, and the boundary operator is defined by counting with sign the number of gradient flow lines between critical points of relative index one. The Morse Homology Theorem says that on a closed finite dimensional smooth manifold the Morse homology with integer coefficients is isomorphic to the singular homology of the manifold with integer coefficients, cf. [8, 54, 78].

In more detail, a function $f : M \to \mathbb{R}$ is called a **Morse** function if all of its critical points are **nondegenerate**, i.e. the **Hessian** of f at every critical point is nondegenerate, cf. Definition 3.1 of [8] or [53]. The assumption that $p \in Cr(f)$ is a nondegenerate critical point implies that the **stable manifold** $W^s(p)$ and the **unstable manifold** $W^u(p)$, defined by

$$W^s(p) = \{x \in M \mid \lim_{t \to \infty} \varphi_t(x) = p\}$$
$$W^u(p) = \{x \in M \mid \lim_{t \to -\infty} \varphi_t(x) = p\},$$

are embedded open disks, where φ_t is the 1-parameter group of diffeomorphisms generated by the negative of the gradient vector field. The **index** of p, defined as the dimension of the subspace of $T_p M$ on which the Hessian is negative definite, coincides with the dimension of the disk $W^u(p)$, cf. Theorem 4.2 of [8].

If all the stable and unstable manifolds of f intersect transversally, i.e.

$$W^u(q) \pitchfork W^s(p)$$

for all $p, q \in Cr(f)$, then the pair (f, g) is called a **Morse-Smale pair**, and f is said to be **Morse-Smale** or to satisfy the **Morse-Smale transversality** condition on (M, g). When (f, g) is a Morse-Smale pair,

$$W(q, p) = W^u(q) \cap W^s(p) \subset M$$

is either empty or an embedded submanifold of dimension $\lambda_q - \lambda_p$, where λ_q and λ_p are the indices of q and p respectively, cf. Proposition 6.2 of [8]. In particular, $W(q, p)$ is a one dimensional manifold when $\lambda_q - \lambda_p = 1$, and $W(q, p)$ consists of the images of finitely many gradient flow lines from q to p when M is compact and $\lambda_q - \lambda_p = 1$, cf. Corollary 6.29 of [8]. Thus, if M is compact and the pair (f, g) is Morse-Smale, then we can count the number of gradient flow lines between any two critical points of relative index one. This is sufficient to define the Morse-Smale-Witten chain complex with coefficients in $\mathbb{Z}_2$, but to define the complex with integer coefficients we need to consider orientations.

Arbitrarily choosing orientations on the unstable manifolds of a Morse-Smale pair (f, g) determines orientations on the normal bundles of the stable manifolds, which then determines signs associated to the gradient flow lines between critical points of relative index one (see Sect. 2.2). With these signs we can define the Morse-Smale-Witten chain complex over $\mathbb{Z}$.

Definition 1.1 (Morse-Smale-Witten Chain Complex) Let $f : M \to \mathbb{R}$ be a smooth Morse-Smale function on a closed smooth Riemannian manifold (M, g) of dimension $m < \infty$. Fix orientations on the unstable manifolds of (f, g). The **Morse-Smale-Witten chain complex with coefficients in** $\mathbb{Z}$ is defined to be the chain complex $(C_*(f), \partial_*)$, where $C_k(f)$ is the free abelian group generated by the critical points of f of index k for all $k = 0, \ldots, m$, and $\partial_k : C_k(f) \to C_{k-1}(f)$ is defined on a generator $q \in C_k(f)$ to be

$$\partial_k(q) = \sum_{p \in Cr_{k-1}(f)} \left(\sum_{v \in \mathcal{M}(q,p)} \epsilon(v) \right) p,$$

where $\epsilon(v) = \pm 1$ denotes the sign associated to the unparameterized gradient flow line $v \in \mathcal{M}(q, p) = W(q, p)/\mathbb{R}$ by the orientations.

Using the structure of the compactified moduli spaces $\overline{\mathcal{M}}(r, p)$, where $\lambda_r - \lambda_p = 2$ one can show that $(\partial_*)^2 = 0$, cf. Lemma 2.13. Thus, $(C_*(f), \partial_*)$ is a chain complex.

The following theorem is well-known and can be proved using many different approaches [8, 37, 54, 66, 73, 78, 96].

Theorem 1.2 (Morse Homology Theorem) *Let $f : M \to \mathbb{R}$ be a smooth Morse-Smale function on a closed finite dimensional smooth Riemannian manifold (M, g). Then the homology of the Morse-Smale-Witten chain complex with coefficients in $\mathbb{Z}$ is isomorphic to the singular homology of M with coefficients in $\mathbb{Z}$.*

1.2 Extensions and Applications

Morse homology can be extended in several ways. For instance, it can be extended to infinite dimensional manifolds, it can be extended to include critical submanifolds [9], it can be extended to manifolds with boundaries or corners [49], and it can be extended to include local coefficient systems. Moreover, these extensions may overlap, e.g. one can consider infinite dimensional versions of Morse homology with local coefficients or Morse homology on manifolds with boundaries or corners with local coefficients.

In previous work, the first two authors explored various ways to extend Morse homology to include critical submanifolds [9, 10, 44]. In this book we discuss how to extend Morse homology to include local coefficient systems, which we refer to as **twisted Morse homology**. The discussion is limited to Morse functions on closed finite dimensional smooth manifolds. Morse-Bott functions and Morse theory on manifolds with boundaries or corners are not discussed outside of this introduction. Even so, there are several interesting applications of Morse homology with local coefficients that fit within the scope of this work.

For instance, an isomorphism between η-twisted Morse cohomology and **Lichnerowicz cohomology** is proved in Chap. 5, and we show how to use twisted Morse cohomology to compute the Lichnerowicz cohomology of a surface with respect to any closed 1-form on the surface in Sect. 6.1. In Sect. 6.2 we discuss, following a result due to Albers et al. [2], how Morse homology with local coefficients in certain rank one R-modules serves as an obstruction to a manifold having an **associative H-space structure**, and in Sect. 6.3 we discuss how the **Novikov homology** of a closed 1-form can be computed using a Morse complex with coefficients in a local system whose fiber is a Novikov ring. Applying this result, we use twisted Morse complexes to explicitly compute the **Novikov numbers** of closed 1-forms on certain manifolds.

As for other applications, there are at least two areas of mathematics that use Morse homology with local coefficients on manifolds with boundaries or corners which should be mentioned. Although this book does not discuss twisted Morse homology on manifolds with boundaries or corners outside of this introduction, the results, techniques, and proofs used in this book should extend to that context with the right definitions and assumptions.

The first area that deserves mention is **Floer homology** for 3 and 4-manifolds, and in particular, the approach based on the Seiberg-Witten monopole equations [47]. In fact, Section 2.4 of [47] outlines an approach for extending Morse homology to manifolds with boundary where the gradient of the function is tangent to the boundary, and Section 2.7 of [47] gives an outline of how to extend Morse homology to include local coefficients. Kronheimer and Mrowka do not include proofs in Section 2.7 of their book because, "In the main part of this book, the Floer homology of a 3-manifold will be constructed by taking these constructions of Morse theory and repeating them in an infinite dimensional setting." [47, p. 1] We note that

Theorems 3.9 and 4.1 of this book prove the claims contained in Section 2.7 of [47] within the context of closed finite dimensional smooth manifolds.

The second area that deserves mention is **symplectic cohomology**, and more specifically, the proof of Viterbo's Theorem given by Abouzaid [1]. Viterbo's Theorem asserts that there is an isomorphism between the twisted homology of the free loop space $\mathcal{L}Q$ of a closed differentiable manifold Q and the symplectic cohomology of its cotangent bundle T^*Q. The proof given by Abouzaid models the free loop space $\mathcal{L}Q$ as a direct limit of spaces of piecewise geodesics $\mathcal{L}_{gr}^r Q$, which are finite dimensional smooth manifolds with corners. The twisted Morse homology of Morse functions on the finite dimensional approximations $\mathcal{L}_{gr}^r Q$ whose gradient flows point outward at the boundary is used in Chapter 11 of [1]. In fact, Proposition 3.13 in Chapter 11 of [1] would be a corollary of Theorems 3.9 and 4.1 of this book extended to Morse-Smale functions on finite dimensional smooth manifolds with corners whose gradient flows point outward at the boundary.

1.3 Summary of Main Results

We now outline the main results and methods included in this book.

A **local coefficient system** on a topological space X assigns a fiber G, which can be a group, ring, module, or field, to every point in the space, and to every homotopy class of paths rel endpoints it assigns a homomorphism in the appropriate category between the fibers of the endpoints. We take as our starting point a **bundle of abelian groups**, which is a local coefficient system in the category of abelian groups (Definition 2.1). Most of the results in Chaps. 2–4 are proved for bundles of abelian groups, but the proofs carry over without change to bundles of rings, modules, and fields (Remark 2.7). A bundle of abelian groups G over a pointed space (X, x_0) induces a representation

$$\pi_1(X, x_0) \wr G_{x_0} \to G_{x_0}$$

on the fiber G_{x_0} over the basepoint x_0, which can be viewed as an isomorphism class of local coefficient systems (Theorem 2.3).

The definition of the boundary operator ∂_*^G in the twisted Morse-Smale-Witten chain complex depends on the homomorphisms associated to the gradient flow lines by the local coefficient system (Definition 2.11), but the homology of the twisted Morse-Smale-Witten chain complex only depends on the isomorphism class of the local coefficient system (Theorem 3.9). The distinction between a bundle of abelian groups and its isomorphism class is examined in detail in Sect. 2.4 with Examples 2.14 and 2.15, where we compute the twisted Morse homology of S^1 and $\mathbb{R}P^2$ with respect to several different local coefficient systems.

In Sect. 2.6 we prove the (untwisted) **Morse Homology Theorem for a Covering Space** (Theorem 2.20) and the **Morse Eilenberg Theorem** (Theorem 2.21). The Morse Eilenberg Theorem relates the twisted Morse homology of a Morse-

1.3 Summary of Main Results

Smale pair $(f, \mathfrak{g})$ with coefficients in a bundle of abelian groups G to the homology of the Morse-Smale-Witten chain complex of $(f, \mathfrak{g})$ pulled back to the universal cover $\widetilde{M}$ of the manifold and twisted by the representation determined by the local coefficient system

$$\pi_1(M, x_0) \times G_{x_0} \to G_{x_0}$$

and the action of $\pi_1(M, x_0)$ on the universal cover $\widetilde{M}$ by deck transformations. One consequence of the Morse Eilenberg Theorem is that the twisted Morse homology of a Morse-Smale pair $(f, \mathfrak{g})$ can be computed using the representation determined by the local coefficient system, instead of the homomorphisms associated to the gradient flow lines by the local coefficient system. This shows that the twisted Morse homology of a Morse-Smale pair $(f, \mathfrak{g})$ only depends on the isomorphism class of G, a fact that is proved separately using an entirely different approach in the next chapter. Remark 2.23 discusses how one might use the Morse Eilenberg Theorem (Theorem 2.21) and results from Conley index theory or Novikov theory (discussed in Sect. 6.3) to prove the Twisted Morse Homology Theorem (Theorem 4.1).

Chapter 3 is devoted to proving an **Invariance Theorem** (Theorem 3.9), which shows that the homology of the Morse-Smale-Witten chain complex of a Morse-Smale pair $(f, \mathfrak{g})$ with coefficients in a bundle of abelian groups G depends only on the isomorphism class of G and not at all on the Morse-Smale pair $(f, \mathfrak{g})$. The proof of Theorem 3.9 is fairly standard within Morse theory and Floer theory, except for the necessary additions required to take into account local coefficient systems. Theorem 3.9 allows us to pick any Morse-Smale pair $(f, \mathfrak{g})$ when computing twisted Morse homology, a fact that we use throughout the rest of the book.

Chapter 4 is devoted to proving the **Twisted Morse Homology Theorem** (Theorem 4.1), which says that the homology of the twisted Morse-Smale-Witten chain complex of $(f; G)$ is isomorphic to the singular homology of M with coefficients in G. In [8] the first two authors presented a proof of the (untwisted) Morse Homology Theorem using **Conley index theory** following [73]. The Conley index theory approach uses a filtration of index pairs associated to the gradient flow which has properties similar to a CW filtration. So, with the Conley index theory approach it is unnecessary to address the question of whether or not the unstable manifolds of a Morse-Smale function determine a CW-structure—a deep result that is usually proved with additional assumptions on the Riemannian metric (see the beginning of Sect. 4.4).

Rather than extending the Conley index theory approach to the Morse Homology Theorem to include local coefficients, we choose instead in this book to work directly with the CW-structures determined by certain Morse-Smale pairs $(f, \mathfrak{g})$. The Invariance Theorem (Theorem 3.9) shows that there is no loss of generality to the Twisted Morse Homology Theorem (Theorem 4.1) if it is proved for a restricted class of Morse-Smale pairs $(f, \mathfrak{g})$, as long as at least one pair $(f, \mathfrak{g})$ in the restricted class exists. The main advantage to working with CW-structures, instead of Conley index filtrations, is that one can directly compare the twisted Morse-Smale-Witten boundary operator to **Steenrod's cellular boundary operator** with local coefficients (Lemma 4.10). This approach also illuminates certain technical issues that arise when extending CW-homology to include local coefficient systems. Specifically, Steenrod's CW-chain complex with local coefficients is not defined for all CW-complexes. Instead, one must restrict to a subcategory of CW-complexes where Steenrod's cellular boundary operator is well-defined [83]. We choose to restrict to the category of **regular CW-complexes**, which are CW-complexes where the attaching maps are homeomorphisms (Definition 4.4).

Regular CW-complexes are much more restrictive than general CW-complexes. For instance, the CW-structure determined by the usual height function on S^1 is not regular (Example 2.14), although it is possible to deform S^1 so that the unstable manifolds of a height function do determine a regular CW-structure (Example 4.9). Similarly, the unstable manifolds of a tilted height function on a torus do not determine a regular CW-structure, cf. Example 7.11 of [8]. However, on every closed finite dimensional smooth manifold M it is always possible to find a Morse-Smale pair (f, g) whose unstable manifolds determine a regular CW-structure on M. We prove this result by constructing a Morse-Smale pair (f, g) whose unstable manifolds coincide with a smooth triangulation on M (Theorem 4.12). We expect that this result should be of independent interest in **combinatorial Morse theory** (Remark 4.15).

The isomorphism between the twisted Morse homology of a Morse-Smale pair (f, g) whose unstable manifolds determine a regular CW-structure and the homology of Steenrod's CW-chain complex with local coefficients is proved in Lemma 4.10. We also include for completeness a detailed proof of the fact that the homology of Steenrod's CW-chain complex with local coefficients is isomorphic to singular homology with local coefficients (Lemma 4.8). These two lemmas, together with the Invariance Theorem (Theorem 3.9) and Theorem 4.12, constitute our proof of the Twisted Morse Homology Theorem (Theorem 4.1).

A key step in the proof of Lemma 4.10 involves relating the signed count of the number of gradient flow lines between two critical points of relative index one to the degree of an attaching map in the associated CW-complex, i.e. $\#\mathcal{M}(q, p) = [e_q^k : e_p^{k-1}]$. This result follows for Morse-Smale pairs (f, g) (with possibly some additional assumptions on g) using the manifolds with corners structure on compactified moduli spaces of gradient flow lines (Remark 4.11), but it is also easy to see directly for the specific Morse-Smale pair (f, g) we construct in Theorem 4.12 (Remark 4.16). Thus, the proof of the Twisted Morse Homology Theorem (Theorem 4.1) given in Chap. 4 is mostly independent of the manifolds with corners structure on the compactified moduli spaces; although the proof of the Invariance Theorem (Theorem 3.9) in Chap. 3 does rely heavily on the manifolds with corners structure on compactified moduli spaces of gradient flow lines.

In Chap. 5 we examine some aspects of **twisted Morse cohomology**. After defining a general Morse-Smale-Witten cochain complex with coefficients in a bundle of abelian groups G (Definition 5.1), we quickly limit our discussion to the local coefficient system $G = e^\eta$ defined by a closed 1-form (Example 2.5), i.e. to an η-twisted Morse-Smale-Witten cochain complex (Definition 5.6). A closed 1-form η on a smooth manifold M can be used both to twist the Morse-Smale-Witten

1.3 Summary of Main Results

coboundary operator and to deform the differential in the de Rham complex via

$$d_\eta \xi = d\xi + \eta \wedge \xi,$$

where $\xi \in \Omega^*(M, \mathbb{R})$. The resulting cohomology groups are known as the **Lichnerowicz cohomology** groups $H^*_\eta(M)$ (Sect. 5.2).

Lichnerowicz cohomology is an invariant used to study **locally conformal symplectic (LCS)** manifolds. An LCS manifold is a smooth manifold with a smooth nondegenerate 2-form Ω which becomes closed locally when multiplied by a smooth positive function (Definition 5.9). Associated to every LCS form Ω there is a closed 1-form η, known as the **Lee form**, which satisfies the equation

$$d\Omega = -\eta \wedge \Omega$$

(Proposition 5.10). The de Rham cohomology class of the Lee form η only depends on the conformal class of the LCS form Ω (Proposition 5.12), and the η-twisted Morse homology and the η-twisted Morse cohomology only depend on the de Rham cohomology class of η (Corollaries 3.10 and 5.7). Thus, the η-twisted Morse homology and cohomology groups are invariants of the conformal class of a locally conformal symplectic form Ω with associated Lee form η (Corollary 5.13).

The main result in Chap. 5 is the η-**Twisted Morse de Rham Theorem** (Theorem 5.22), which says that the η-twisted Morse cohomology groups are isomorphic to the Lichnerowicz cohomology groups defined by $-\eta$. Our proof of Theorem 5.22 relies on the η-**twisted Morse Poincaré Lemma** (Lemma 5.21) and is modeled on the proof of the de Rham Theorem found in Section V.9 of [17]. We conclude our discussion of twisted Morse cohomology in Sect. 5.4, where we discuss its relationship with **sheaf cohomology**. In particular, we discuss a result due to Vaisman [86] that says that the Lichnerowicz cohomology groups $H^*_{-\eta}(M)$ are isomorphic to the cohomology of the manifold M with coefficients in a sheaf $\mathcal{F}_\eta(M)$ (Theorem 5.28).

The isomorphisms proved in Chaps. 2–5 show that homology with local coefficients and Lichnerowicz cohomology on closed finite dimensional smooth manifolds can be computed using finitely generated chain and cochain complexes. This is enough to establish the **invariance of the twisted Euler number** for homology with local coefficients in a bundle of finitely generated free R-modules (Theorem 4.21) and Lichnerowicz cohomology (Corollary 5.25) using the **Euler-Poincaré Theorem** (Theorem 4.19) for modules over a principle ideal domain R. Moreover, twisted Morse complexes and cochain complexes often have fewer generators than Steenrod's CW-chain complex for regular CW-complexes (compare Examples 2.14 and 4.9). Thus, twisted Morse chain and cochain complexes are superb tools for computing homology and cohomology with local coefficients, a topic we turn to in Chap. 6.

We begin Chap. 6 by reviewing a result due to León et al. [25] that says that if a nonzero closed 1-form η on a closed smooth manifold M is parallel with respect to some Riemannian metric on M, then the Lichnerowicz cohomology $H^*_\eta(M)$ vanishes (Theorem 6.1). Thus, the η-twisted Morse cohomology groups serve as obstructions to the existence of **nonzero closed parallel 1-forms** on closed smooth manifolds (Corollary 6.8), as well as invariants for the conformal class of an LCS form (Corollary 5.13). This shows why it might be useful to compute η-twisted Morse cohomology groups, and in Example 6.12 we use twisted Morse cochain complexes to explicitly compute the Lichnerowicz cohomology of a surface S of genus two with respect to all closed 1-forms η on the surface. The invariance of the η-twisted Euler number and the small number of generators required for the twisted Morse cochain complex provide strong constraints on the solution. In fact, these constraints imply that there are only 4 possibilities for the Lichnerowicz cohomology groups $H^*_\eta(S)$, no matter which closed 1-form $\eta \in \Omega^1_{cl}(S, \mathbb{R})$ is considered, and our computation shows that 2 of these 4 possibilities actually occur. The approach used in Example 6.12 should extend to many other surfaces and manifolds that have "conducive" CW-structures.

In Sect. 6.2 we present a proof of a result due to Albers, Frauenfelder, and Oancea [2] that says that if X is a path connected **associative H-space** with a local coefficient system $\mathcal{L}$ of rank one R-modules satisfying certain conditions, then the singular homology of X with coefficients in $\mathcal{L}$ vanishes (Proposition 6.14). Thus, the Twisted Morse Homology Theorem (Theorem 4.1) implies that Morse homology with coefficients in certain local coefficient systems serve as obstructions to a closed smooth manifold M having an associative H-space structure (Corollary 6.17). Moreover, the invariance of the twisted Euler number (Corollary 4.22) implies that the Euler number serves as an obstruction to the existence of an associative H-space structure on a closed smooth manifold M with $H^1_{dR}(M; \mathbb{R}) \neq 0$ (Corollary 6.18). In Examples 6.19–6.22 we consider the twisted Morse homology of various manifolds with coefficients in local systems $\mathcal{L}$ meeting the conditions listed in Proposition 6.14. In particular, we use twisted Morse complexes to shows that a surface of genus two is not an associative H-space, and $\mathbb{R}P^n$ is not an associative H-space when n is even, cf. Example 1 of [2].

Section 6.3 contains an exposition of Novikov homology, which extends Morse homology from exact 1-forms to closed 1-forms, and its relationship with twisted Morse homology. Novikov homology is more general than (untwisted) Morse homology; however, the Novikov Principle implies that Novikov homology can be computed using twisted Morse homology (Corollary 6.32). The Novikov chain complex is constructed by pulling back a closed one form $\zeta \in \Omega^1_{cl}(M, \mathbb{R})$ and a generic Riemannian metric g on M to a covering space where ζ becomes exact. The flow of the pullback of the pair (ζ, g) to a covering space where ζ is exact will be a Morse-Smale gradient flow. However, if the **rank** of ζ is greater than zero, then the pullback of ζ will have an infinite number of zeros (Corollary 6.26).

To account for the infinite number of zeros introduced when pulling back the form to a covering space the **Novikov ring** (Definition 6.27), which consists of countable "half infinite" Laurent series with integer coefficients and exponents in $\mathbb{R}$, is used. In Definition 6.28 we introduce a local coefficient system $\mathcal{L}_\zeta$ of rank one Nov-modules associated to a closed 1-form ζ. The local coefficient system $\mathcal{L}_\zeta$ is

1.3 Summary of Main Results

suitable for use with twisted Morse homology, and it is in the isomorphism class of local coefficient systems commonly used for Novikov homology (Proposition 6.29).

Using twisted Morse homology with coefficients in the local system $\mathcal{L}_\zeta$ we compute the **Novikov numbers** (Definition 6.35) of all closed 1-forms on various manifolds (Examples 6.39–6.42). The Novikov numbers $b_k([\zeta])$ and $q_k([\zeta])$ for $k = 0, \ldots, m$, which generalize the Betti numbers and the torsion numbers of a manifold (Proposition 6.36), satisfy the **Novikov inequalities** (Theorem 6.37), which generalize the Morse inequalities, cf. Theorem 3.33 of [8]. Hence, the Novikov inequalities give lower bounds on the number of zeros of a closed Morse 1-form ζ. In Example 6.40 we note that the local system of rank one Nov-modules $\mathcal{L}_\zeta$ satisfies the conditions listed in Proposition 6.14 whenever ζ is a non exact closed 1-form, and since a torus is an associative H-space this implies that the Novikov numbers of a non exact closed 1-form on a torus vanish. Our last example is an example where the Novikov inequalities give a nontrivial lower bound on the number of zeros of a non exact closed 1-form. In Example 6.42 we use a twisted Morse complex with coefficients in the local system $\mathcal{L}_\zeta$ to compute the Novikov numbers of a surface S of genus two with respect to any closed 1-form $\zeta \in \Omega^1_{cl}(S, \mathbb{R})$. The computation shows that every non exact closed Morse 1-form on S must have at least two zeros with Morse index 1.

Chapter 2
The Morse Complex with Local Coefficients

In this chapter the Morse-Smale-Witten chain complex with coefficients in a bundle of abelian groups over a finite dimensional smooth Riemannian manifold is constructed. The twisted Morse chain complex is described in detail for S^1 and real projective space. Relationships with Morse complexes on covering spaces are discussed, and a Morse theoretic version of Eilenberg's Theorem is proved.

2.1 Local Coefficients

A bundle of abelian groups G over a topological space X is a functor from the fundamental groupoid of X to the category of abelian groups. More explicitly, we have the following.

Definition 2.1 A **bundle of abelian groups** G over a topological space X associates to every point $x \in X$ an abelian group G_x and to every continuous path $\gamma : [0, 1] \to X$ a homomorphism $\gamma_* : G_{\gamma(1)} \to G_{\gamma(0)}$ such that the following conditions are satisfied.

1. If two paths $\gamma_1, \gamma_2 : [0, 1] \to X$ from $x \in X$ to $y \in X$ are homotopic rel endpoints, then the homomorphisms from G_y to G_x associated to γ_1 and γ_2 are the same, i.e. $(\gamma_1)_* = (\gamma_2)_*$.
2. If $\gamma : [0, 1] \to X$ is constant, then γ_* is the identity.
3. If $\gamma_1, \gamma_2 : [0, 1] \to X$ are paths with $\gamma_1(1) = \gamma_2(0)$, then $(\gamma_1 \gamma_2)_* = (\gamma_1)_* \circ (\gamma_2)_*$, where $\gamma_1 \gamma_2$ denotes the concatenation of γ_1 and γ_2.

Letting $\gamma_2(t) = \gamma_1(1-t)$ in (3), we see that the above conditions imply that the homomorphism γ_* associated to a path γ is in fact an isomorphism.

Note If G is any abelian group and γ_* is the identity map for all paths $\gamma : [0, 1] \to X$, then associating $G = G_x$ to every point $x \in X$ determines a **constant** bundle of abelian groups.

Definition 2.2 Suppose that G_1 and G_2 are both bundles of abelian groups over a topological space X. If there exists a family of isomorphisms $\Phi : G_1 \to G_2$ such that for every continuous path $\gamma : [0, 1] \to X$ the diagram

$$\begin{array}{ccc} (G_1)_{\gamma(1)} & \xrightarrow{\gamma_*^{G_1}} & (G_1)_{\gamma(0)} \\ \Phi_{\gamma(1)} \downarrow & & \downarrow \Phi_{\gamma(0)} \\ (G_2)_{\gamma(1)} & \xrightarrow{\gamma_*^{G_2}} & (G_2)_{\gamma(0)} \end{array}$$

commutes, then G_1 and G_2 are said to be **isomorphic**.

Note A bundle of abelian groups that is isomorphic to a constant bundle is called **simple**. A bundle of abelian groups G is simple if and only if for any $x, y \in X$ the homomorphism γ_* is independent of the path γ from x to y.

A bundle of abelian groups G over a pointed space (X, x_0) induces a representation

$$\pi_1(X, x_0) \times G_{x_0} \to G_{x_0},$$

i.e. $[\gamma] \in \pi_1(X, x_0)$ determines an isomorphism $\gamma_* : G_{x_0} \to G_{x_0}$ and $(\gamma_1 \gamma_2)_* = (\gamma_1)_* \circ (\gamma_2)_*$ for any $[\gamma_1], [\gamma_2] \in \pi_1(X, x_0)$. The following converse is well known, cf. Theorem VI.1.11 and VI.1.12 of [93].

Theorem 2.3 *Let X be a nonempty path connected topological space with a basepoint x_0, and let G_0 be an abelian group on which $\pi_1(X, x_0)$ operates. Then there exists a bundle of abelian groups G over X with $G_{x_0} = G_0$ which induces the operation of $\pi_1(X, x_0)$ on G_0, and the bundle G is unique up to isomorphism.*

Remark 2.4 The boundary operator in the twisted Morse-Smale-Witten chain complex (Definition 2.11) depends on the choice of the bundle of abelian groups G. However, we will show that the homology of the twisted Morse-Smale-Witten complex only depends on the isomorphism class of G, i.e. only on the representation $\pi_1(X, x_0) \times G_{x_0} \to G_{x_0}$ (Theorems 2.21 and 3.9).

Example 2.5 (The Local Coefficient System e^η Determined by a Closed 1-Form) Let $\eta \in \Omega^1_{cl}(M, \mathbb{R})$ be a closed smooth real valued 1-form on a finite dimensional smooth manifold M. To each point $x \in M$ associate the additive abelian group $\mathbb{R}$, and to each smooth path $\gamma : [0,1] \to M$ associate the homomorphism $\gamma_* : \mathbb{R}_{\gamma(1)} \to \mathbb{R}_{\gamma(0)}$ defined by

$$\gamma_*(s) = e^{\int_1^0 \gamma^*(\eta)} \cdot s \quad \text{for all } s \in \mathbb{R}.$$

Since every continuous path in M is homotopic rel endpoints to a smooth path, Stokes' Theorem shows that this defines a bundle of (additive) $\mathbb{R}$ groups over M. The above definition of γ_* extends to paths $\gamma : \overline{\mathbb{R}} \to M$ using any diffeomorphism $\overline{\mathbb{R}} \approx [0,1]$. We will denote this flat line bundle by e^η.

Proposition 2.6 *If $\eta_1, \eta_2 \in \Omega^1_{cl}(M, \mathbb{R})$ are in the same de Rham cohomology class, then e^{η_1} is isomorphic to e^{η_2}.*

Proof By assumption there exists a smooth function $h : M \to \mathbb{R}$ with $\eta_1 - \eta_2 = dh$. Define a family of isomorphisms $\Phi : e^{\eta_1} \to e^{\eta_2}$ by $\Phi_x(s) = e^{-h(x)} \cdot s$ for all $x \in M$ and $s \in \mathbb{R}$. Then the following diagram commutes for any path $\gamma : [0,1] \to \mathbb{R}$

$$\begin{array}{ccc} \mathbb{R} & \xrightarrow{\times e^{\int_1^0 \gamma^*(\eta_1)}} & \mathbb{R} \\ {\scriptstyle \times e^{-h(\gamma(1))}} \downarrow & & \downarrow {\scriptstyle \times e^{-h(\gamma(0))}} \\ \mathbb{R} & \xrightarrow{\times e^{\int_1^0 \gamma^*(\eta_2)}} & \mathbb{R} \end{array}$$

because $e^{\int_1^0 \gamma^*(\eta_1)} = e^{\int_1^0 \gamma^*(\eta_2 + dh)} = e^{\int_1^0 \gamma^*(\eta_2)} e^{h(\gamma(0)) - h(\gamma(1))}$. □

Remark 2.7 Note that e^η is not only a bundle of abelian groups with respect to addition on the fibers, but it is also a bundle of commutative rings, a rank one bundle of $\mathbb{R}$-modules, and a flat $\mathbb{R}$-vector bundle of dimension one, also known as a flat line bundle. That is, the homomorphisms $\gamma_*(s) = e^{\int_1^0 \gamma^*(\eta)} \cdot s$ are isomorphisms in each of those categories. Moreover, the notion of isomorphic bundles from Definition 2.2 carries over to those other categories by requiring Φ to be an isomorphism in the category under consideration, and the proof of Proposition 2.6 carries over to those other categories without modification.

Remark 2.8 We can also define a local coefficient system a^η for any $a > 0$ by replacing $e^{\int_1^0 \gamma^*(\eta)}$ with $a^{\int_1^0 \gamma^*(\eta)} = e^{(\ln a) \int_1^0 \gamma^*(\eta)}$ in the preceding example. The proof of Proposition 2.6 holds for the system a^η.

2.2 Path Components of Compactified Moduli Spaces

Let $f : M \to \mathbb{R}$ be a smooth Morse-Smale function on a closed finite dimensional smooth Riemannian manifold M. Let $Cr(f) = \{p \in M \mid df_p = 0\}$ denote the set of critical points of f, let $Cr_k(f) \subset Cr(f)$ denote the set of critical points of index

k, and for any $p \in Cr(f)$ let λ_p denote the index of p. For $p, q \in Cr(f)$ let $W^u(q) \subset M$ be the unstable manifold of q, $W^s(p) \subset M$ the stable manifold of p, and define

$$W(q, p) = W^u(q) \cap W^s(p) \subset M.$$

If this space is nonempty, then one says that q is succeeded by p, i.e. $q \succeq p$. In this case, $W(q, p)$ is a noncompact smooth manifold of dimension $\lambda_q - \lambda_p$. Choosing orientations for the unstable manifolds $W^u(q)$ for all $q \in Cr(f)$ determines an orientation on $W(q, p)$ for all $p, q \in Cr(f)$ via the short exact sequence

$$0 \longrightarrow T_*W(q, p) \hookrightarrow T_*W^u(q)|_{W(q,p)} \longrightarrow \nu_*(W(q, p), W^u(q))|_{W(q,p)} \longrightarrow 0$$

where the fibers of the normal bundle are isomorphic to $T_p W^u(p)$ via the gradient flow. Taking a quotient by the action of $\mathbb{R}$ given by the flow of $-\nabla f$ then gives a smooth manifold

$$\mathcal{M}(q, p) = W(q, p)/\mathbb{R}$$

of dimension $\lambda_q - \lambda_p - 1$ [46], which we orient as follows. For any regular value y between $f(p)$ and $f(q)$ we identify $\mathcal{M}(q, p) = W(q, p) \cap f^{-1}(y)$, and for any $x \in W(q, p) \cap f^{-1}(y)$ we declare B_x to be a positive basis for $T_x \mathcal{M}(q, p)$ if and only if $(-(\nabla f)(x), B_x)$ is a positive basis for $T_x W(q, p)$. (See Section 6.1 of [66] or Proposition 3.10 of [90] for more details.)

The moduli space $\mathcal{M}(q, p)$ has a compactification $\overline{\mathcal{M}}(q, p)$ consisting of the piecewise gradient flow lines from q to p, which can be given the structure of a smooth manifold with corners [18, 19, 48, 66, 67, 91]. As in Section 6.1 of [66], we orient the (codimension) 1-stratum using the convention that an outward pointing normal vector field followed by a positive basis for a tangent space of $\partial^1 \overline{\mathcal{M}}(q, p)$ should be a positive basis for a tangent space of $\overline{\mathcal{M}}(q, p)$.

A piecewise gradient flow line from q to p can be identified with its image in M, which is an element of $\mathcal{P}^c(M)$, the space of all nonempty closed subsets of M with the Hausdorff topology. This identification is compatible with the topology of the smooth manifold with corners $\overline{\mathcal{M}}(q, p)$ in the sense that the map that sends an element of $v \in \overline{\mathcal{M}}(q, p)$ to its image $Im(v)$ is a homeomorphism onto its image $Im(\overline{\mathcal{M}}(q, p))$ in $\mathcal{P}^c(M)$ [10, 43, 91].

Denote the path component of $v \in \overline{\mathcal{M}}(q, p)$ by $\overline{\mathcal{M}}(q, p; [v])$. Let $(\gamma_1, \ldots, \gamma_l)$ be a sequence of gradient flow lines with

$$\lim_{t \to -\infty} \gamma_1(t) = q,$$

$$\lim_{t \to \infty} \gamma_j(t) = \lim_{t \to -\infty} \gamma_{j+1}(t) \text{ for all } j = 1, \ldots, l-1,$$

$$\lim_{t \to \infty} \gamma_l(t) = p,$$

2.2 Path Components of Compactified Moduli Spaces

and denote the corresponding sequence of unparameterized gradient flow lines by $(v_1, \ldots, v_l)$. We will write $[(v_1, \ldots, v_l)] = [v]$ to indicate that the image of the piecewise gradient flow line $(v_1, \ldots, v_l)$

$$Im(v_1, \ldots, v_l) = Im(\gamma_1, \ldots, \gamma_l) = \bigcup_{j=1}^{l} \gamma_j(\mathbb{R}) \in \mathcal{P}^c(M)$$

is in the same path component as $Im(v)$ in $\mathcal{P}^c(M)$.

Lemma 2.9 *Let $r, p \in Cr(f)$. If $v \in \mathcal{M}(r, p)$, then the closure of $\mathcal{M}(r, p; [v])$ in $\overline{\mathcal{M}}(r, p)$ consists of the piecewise gradient flow lines from r to p that are in the same path component as v. Moreover, when $\lambda_r - \lambda_p = 2$ we have*

$$\partial^1 \overline{\mathcal{M}}(r, p; [v]) = \partial \overline{\mathcal{M}}(r, p; [v]) = (-1) \bigcup_{\substack{r \geq q \geq p \\ [v]=[(v_1, v_2)]}} \mathcal{M}(r, q; [v_1]) \times \mathcal{M}(q, p; [v_2])$$

as oriented manifolds. Thus when $\lambda_r - \lambda_p = 2$,

$$\sum_{r \geq q \geq p} \sum_{\substack{[v]=[(v_1, v_2)] \\ (v_1, v_2) \in \mathcal{M}(r,q) \times \mathcal{M}(q,p)}} \epsilon(v_1) \epsilon(v_2) = 0$$

where $\epsilon(v_j) = \pm 1$ is the sign of the zero dimensional oriented manifold v_j for $j = 1, 2$.

Proof The boundary of the smooth manifold with corners $\overline{\mathcal{M}}(r, p)$ consists of the broken gradient flow lines from r to p, and the interior $\mathcal{M}(r, p)$ consists of the (nonbroken) gradient flow lines from r to p. Thus, the closure of the path component $\mathcal{M}(r, p; [v])$ consists of the piecewise gradient flow lines in $\overline{\mathcal{M}}(r, p)$ that are in same path component as v.

The second statement follows immediately from Theorem 3.6 of [66] and Theorem 8.1 of [67]. The last statement follows because $\overline{\mathcal{M}}(r, p; [v])$ is a compact connected oriented 1-dimensional smooth manifold with boundary when $\lambda_r - \lambda_p = 2$ (thus diffeomorphic to S^1 or $[0, 1]$), and the oriented sum of the signs associated to the boundary points of a 1-dimensional compact smooth manifold is zero. □

Remark 2.10 Formulas similar to those in the preceding lemma can be found in Lemma 3.4 of [5], Proposition 5.2 of [19], Theorem 2 of [18], Sections 2.14 and 2.15 of [48], Lemma 4.3 of [78], and Proposition 4.4 of [90]. However, we will following the sign conventions in [66] and [67].

2.3 Twisting the Morse-Smale-Witten Boundary Operator

Let G be a bundle of abelian groups over M and let $\gamma^\nu : [0,1] \to M$ be a continuous path from p to q whose image coincides with the image of some element $\nu \in \overline{\mathcal{M}}(q,p;[\tilde{\nu}])$, where $q, p \in Cr(f)$ and $\tilde{\nu} \in \overline{\mathcal{M}}(q,p)$. Lemma 2.9 implies that any two paths representing elements of a path component $\overline{\mathcal{M}}(q,p;[\tilde{\nu}])$ are homotopic rel endpoints, and hence condition (1) of Definition 2.1 implies that there is a well-defined homomorphism $\gamma_*^\nu : G_q \to G_p$ which is independent of the element $\nu \in \overline{\mathcal{M}}(q,p;[\tilde{\nu}])$.

Definition 2.11 (Twisted Morse-Smale-Witten Chain Complex) Let $f : M \to \mathbb{R}$ be a smooth Morse-Smale function on a closed smooth Riemannian manifold (M, g) of dimension $m < \infty$. Fix orientations on the unstable manifolds of (f, g), and let G be a bundle of abelian groups over M. The **Morse-Smale-Witten chain complex with coefficients in** G is defined to be the chain complex $(C_*(f;G), \partial_*^G)$ where

$$C_k(f;G) \stackrel{def}{=} \left\{ \sum_{q \in Cr_k(f)} gq \;\middle|\; g \in G_q \right\} \approx \bigoplus_{q \in Cr_k(f)} G_q$$

for all $k = 0, \ldots, m$, and the homomorphism $\partial_k^G : C_k(f;G) \to C_{k-1}(f;G)$ is defined on an elementary chain $gq \in C_k(f;G)$ to be

$$\partial_k^G(gq) = \sum_{p \in Cr_{k-1}(f)} \sum_{\nu \in \mathcal{M}(q,p)} \epsilon(\nu) \gamma_*^\nu(g) p,$$

where $\gamma^\nu : [0,1] \to M$ is any continuous path from p to q whose image coincides with the image of $\nu \in \mathcal{M}(q,p)$ and $\epsilon(\nu) = \pm 1$ is the sign determined by the orientation on $\mathcal{M}(q,p)$.

Note Since G_p is abelian, there is an action $\mathbb{Z} \times G_p \to G_p$ that sends $-1 \cdot g$ to the inverse of g. Also, if $G = \mathbb{Z}$ is a constant bundle, then $\gamma_*^\nu = id$ for all $p, q \in Cr(f)$ and the above reduces to the Morse-Smale-Witten chain complex with coefficients in $\mathbb{Z}$, cf. Chapter 7 of [8]. We will sometimes drop the G in the notation ∂_*^G when the choice of the bundle of coefficients is clear.

Definition 2.12 (η-Twisted Morse-Smale-Witten Chain Complex) Let $f : M \to \mathbb{R}$ be a smooth Morse-Smale function on a closed finite dimensional smooth Riemannian manifold (M, g). Fix orientations on the unstable manifolds of (f, g), and let $\eta \in \Omega_{cl}^1(M, \mathbb{R})$. The Morse-Smale-Witten chain complex with coefficients in the local system e^η is called the **η-twisted Morse-Smale-Witten chain complex**. In this case, $C_k(f;e^\eta) \approx C_k(f) \otimes \mathbb{R}$, where $C_k(f)$ is the free

2.3 Twisting the Morse-Smale-Witten Boundary Operator

abelian group generated by the critical points q of index k, and the homomorphism $\partial_k^\eta : C_k(f) \otimes \mathbb{R} \to C_{k-1}(f) \otimes \mathbb{R}$ is given on a critical point $q \in Cr_k(f)$ by

$$\partial_k^\eta(q) = \sum_{p \in Cr_{k-1}(f)} \sum_{v \in \mathcal{M}(q,p)} \epsilon(v) \exp\left(\int_{\mathbb{R}} \gamma_v^*(\eta)\right) p,$$

where $\gamma_v : \overline{\mathbb{R}} \to M$ is any gradient flow line from q to p parameterizing $v \in \mathcal{M}(q,p)$ and $\epsilon(v) = \pm 1$ is the sign determined by the orientation on $\mathcal{M}(q,p)$.

Note Example 2.5 shows that the η-twisted Morse-Smale-Witten chain complex is a special case of the general twisted Morse-Smale-Witten chain complex in Definition 2.11.

Lemma 2.13 *The pair* $(C_*(f;G), \partial_*^G)$ *is a chain complex, i.e.* $(\partial_*^G)^2 = 0$.

Proof Let $r \in Cr(f)$ with $\lambda_r = k+1$ for some $k = 1, \ldots, m-1$, where $m = \dim M$. For any $g \in G_r$ we have

$$\partial_k^G(\partial_{k+1}^G(gr)) = \partial_k^G\left(\sum_{q \in Cr_k(f)} \sum_{v_1 \in \mathcal{M}(r,q)} \epsilon(v_1) \gamma_*^{v_1}(g) q\right)$$

$$= \sum_{q \in Cr_k(f)} \sum_{v_1 \in \mathcal{M}(r,q)} \epsilon(v_1) \partial_k^G\left(\gamma_*^{v_1}(g) q\right)$$

$$= \sum_{q \in Cr_k(f)} \sum_{v_1 \in \mathcal{M}(r,q)} \epsilon(v_1) \sum_{p \in Cr_{k-1}(f)} \sum_{v_2 \in \mathcal{M}(q,p)} \epsilon(v_2) \gamma_*^{v_2}(\gamma_*^{v_1}(g)) p$$

$$= \sum_{p \in Cr_{k-1}(f)} \sum_{q \in Cr_k(f)} \sum_{v_1 \in \mathcal{M}(r,q)} \sum_{v_2 \in \mathcal{M}(q,p)} \epsilon(v_1)\epsilon(v_2) \gamma_*^{(v_1,v_2)}(g) p.$$

Now consider the coefficient in front of some fixed $p \in Cr_{k-1}(f)$.

$$\mathrm{coef}(p) = \sum_{q \in Cr_k(f)} \sum_{(v_1,v_2) \in \mathcal{M}(r,q) \times \mathcal{M}(q,p)} \epsilon(v_1)\epsilon(v_2) \gamma_*^{(v_1,v_2)}(g).$$

We can group the terms in the above sum according to the various path components $\overline{\mathcal{M}}(r,p;[v])$ and use the fact that the homomorphism $\gamma_*^{(v_1,v_2)} = \gamma_*^v$ on each path component to get terms of the form

$$\gamma_*^v(g) \sum_{q \in Cr_k(f)} \sum_{\substack{[v]=[(v_1,v_2)] \\ (v_1,v_2) \in \mathcal{M}(r,q) \times \mathcal{M}(q,p)}} \epsilon(v_1)\epsilon(v_2).$$

These terms are zero by Lemma 2.9. □

2.4 Examples

Example 2.14 (A Circle) Consider the height function $f : S^1 \to \mathbb{R}$ on the unit circle $S^1 \subset \mathbb{R}^2$ with a critical point q of index 1 and a critical point p of index 0. Orient the unstable manifold of q clockwise and the unstable manifold of p as $+1$. The (untwisted) Morse-Smale-Witten chain complex of f is

$$\begin{array}{ccccccc} 0 & \longrightarrow & C_1(f) & \xrightarrow{\partial_1} & C_0(f) & \longrightarrow & 0 \\ & & \updownarrow \approx & & \updownarrow \approx & & \\ 0 & \longrightarrow & <q> & \xrightarrow{\partial_1} & <p> & \longrightarrow & 0 \end{array}$$

with $\partial_1(q) = 0$ zero since the two gradient flow lines have opposite orientations. (See Example 7.7 of [8] for more details.)

If G is a bundle of abelian groups over S^1 and $g \in G_q$, then

$$\partial_1^G(gq) = \left(\gamma_*^r(g) - \gamma_*^l(g)\right) p$$

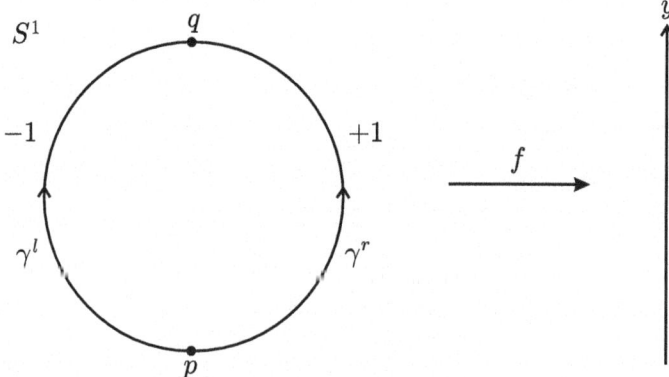

The height function on a circle

where γ^r is a parameterization of the right half of the circle and γ^l is a parameterization of the left half of the circle, both from p to q. Thus, for $k = 0, 1$

$$H_k((C_*(f; G), \partial_*^G)) \approx H_k(S^1; G_q) \text{ if } \gamma_*^r(g) = \gamma_*^l(g) \text{ for all } g \in G_q.$$

However, it is possible to have a bundle of abelian groups over S^1 where $\gamma_*^r \neq \gamma_*^l$. To see this, recall that a representation

$$\pi_1(S^1, q) \times G_q \to G_q$$

2.4 Examples

determines a bundle of abelian groups over S^1 that is unique up to isomorphism. The bundle of abelian groups G is defined by associating the group G_q to every point and arbitrarily fixing a homotopy class of paths rel endpoints $[\gamma_{qx}]$ from q to x for every $x \in S^1$. If γ is a path from $x_0 \in S^1$ to $x_1 \in S^1$ and γ_{qx_0} and γ_{qx_1} are paths from q to x_0 and from q to x_1 in the chosen homotopy classes, then the concatenation $\gamma_{qx_0} \gamma \gamma_{qx_1}^{-1}$ represents an element in $\pi_1(S^1, q)$ and the homomorphism $\gamma_* : G_{x_1} \to G_{x_0}$ is defined to be the associated homomorphism determined by the representation. (For more details see the proof of Theorem VI.1.12 in [93].)

For instance, suppose $G_q = \mathbb{Z}$ and $\pi_1(S^1, q)$ is identified with $\mathbb{Z}$ by sending the clockwise generator to 1. A representation

$$\pi_1(S^1, q) \times \mathbb{Z} \to \mathbb{Z}$$

is then determined by whether $(1, 1) \mapsto 1$ or $(1, 1) \mapsto -1$, since the only group automorphisms of $\mathbb{Z}$ are $\pm\text{id}$. If $(1, 1) \to 1$, then the representation is trivial and the homomorphism associated to any path is the identity. So, assume that $(1, 1) \mapsto -1$, which implies that $(n, g) \mapsto (-1)^n g$ for all $(n, g) \in \pi_1(S^1, q) \times \mathbb{Z}$.

For the chosen homotopy classes of paths $[\gamma_{qx}]$ from q to $x \in S^1$ we will first take the constant path when $x = q$ and choose paths that wrap clockwise n-times around S^1 and then continue clockwise to x whenever $x \neq q$. Taking $x_0 = p$ and $x_1 = q$ in the above we see that $\gamma_*^r(g) = (-1)^n g$ and $\gamma_*^l(g) = (-1)^{n+1} g$, and hence $\partial_1^G(gq) = \left((-1)^n - (-1)^{n+1}\right) gp = 2(-1)^n gp$. Therefore,

$$H_1((C_*(f; G), \partial_*^G)) \approx 0, \quad H_0((C_*(f; G), \partial_*^G)) \approx \mathbb{Z}_2.$$

Alternately, we could choose the constant path when $x = q$ and paths that wrap counterclockwise n-times around S^1 and then continue counterclockwise to $x \in S^1$ when $x \neq q$ for the homotopy classes of paths. Then $\gamma_*^r(g) = (-1)^{-(n+1)} g$ and $\gamma_*^l(g) = (-1)^{-n} g$, and hence $\partial_1^G(gq) = \left((-1)^{-(n+1)} - (-1)^{-n}\right) gp = 2(-1)^{-(n+1)} gp$. The reader can verify that alternate choices for the homotopy class of paths from q to q yield similar boundary operators. Hence, we see that although ∂_* depends on the specific bundle of abelian groups G over S^1, the homology of $(C_*(f; G), \partial_*^G)$ depends only on the isomorphism class of G.

As another example of a bundle of abelian groups over S^1, consider a closed 1-form η on S^1, its associated flat line bundle e^η, and the associated η-twisted Morse-Smale-Witten boundary operator

$$\partial_1^\eta(q) = \left(\exp\left(\int_1^0 (\gamma^r)^*(\eta)\right) - \exp\left(\int_1^0 (\gamma^l)^*(\eta)\right)\right) p.$$

If $\eta = dh$ is exact, then the integral of η along any path from q to p is $h(q) - h(p)$. Hence,

$$\partial_1^\eta(q) = \left(e^{h(q)-h(p)} - e^{h(q)-h(p)}\right)p = 0,$$

and $H_*((C_*(f; \mathbb{R}), \partial_*^\eta)) = H_*(S^1; \mathbb{R})$. However, if η is not exact, then $\int_1^0 (\gamma^r)^*(\eta)$ is not equal to $\int_1^0 (\gamma^l)^*(\eta)$. In this case $\partial_1^\eta(q) \neq 0$, and $H_k((C_*(f; \mathbb{R}), \partial_*^\eta)) = 0$ for all k. Explicitly, consider the form

$$d\theta = \frac{1}{x^2 + y^2}(-y\,dx + x\,dy)$$

and the parameterization of S^1 given by $\gamma(t) = (\cos t, \sin t)$. Then we have

$$\int_1^0 (\gamma^r)^*(d\theta) = \int_{\pi/2}^{-\pi/2} \sin^2 t + \cos^2 t \, dt = -\pi$$

and

$$\int_1^0 (\gamma^l)^*(d\theta) = \int_{\pi/2}^{3\pi/2} \sin^2 t + \cos^2 t \, dt = \pi.$$

Thus, $\partial_1^\eta(q) = (e^{-\pi} - e^\pi)p \neq 0$, and $H_k((C_*(f) \otimes \mathbb{R}, \partial_*^\eta)) \approx 0$ for all k.

Example 2.15 (Real Projective Space) The real projective space $M = \mathbb{R}P^2$ can be viewed as $S^2 \subset \mathbb{R}^3$ with diametrically opposed points identified or as the closed disk $D^2 \subset \mathbb{R}^2$ with diametrically opposed points on the boundary identified. The Morse-Smale function $\tilde{f} : S^2 \to \mathbb{R}$ defined by

$$\tilde{f}(x_1, x_2, x_3) = x_2^2 + 2x_3^2$$

satisfies $\tilde{f}(-x_1, -x_2, -x_3) = \tilde{f}(x_1, x_2, x_3)$, and hence it descends to a Morse-Smale function $f : \mathbb{R}P^2 \to \mathbb{R}$ with three critical points p, q, r of index 0, 1, and 2 respectively, whose unstable manifolds give a CW-structure with three cells e_0, e_1, e_2. (Compare with Example 2.18 and Example 3.7 of [8].) The unstable manifolds, the gradient flow lines, and their orientations are indicated in the following diagram. The convention for the orientations is given by the short exact sequence

$$0 \longrightarrow T_*W(q, p) \hookrightarrow T_*W^u(q)|_{W(q,p)} \longrightarrow \nu_*(W(q, p), W^u(q))|_{W(q,p)} \longrightarrow 0$$

2.4 Examples

where $W(q, p) = W^u(q) \pitchfork W^s(p)$ for any two critical points q and p, and the fibers of the normal bundle are isomorphic to $T_p^u W^u(p)$. The (untwisted) Morse-Smale-Witten chain complex of f is

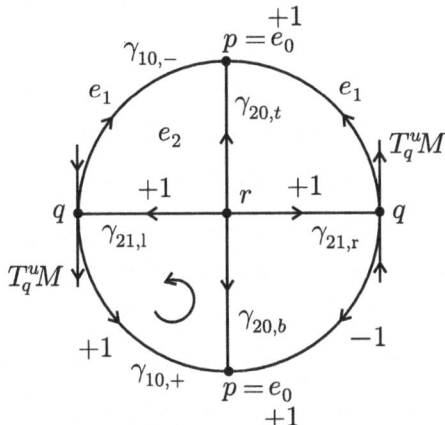

A Morse-Smale function on $\mathbb{R}P^2$

$$0 \longrightarrow C_2(f) \xrightarrow{\partial_2} C_1(f) \xrightarrow{\partial_1} C_0(f) \longrightarrow 0$$

$$0 \longrightarrow <r> \xrightarrow{\partial_2} <q> \xrightarrow{\partial_1} <p> \longrightarrow 0$$

with $\partial_2(r) = 2q$ and $\partial_1(q) = 0$. Thus,

$$H_2((C_*(f), \partial_*)) \approx 0, \quad H_1((C_*(f), \partial_*)) \approx \mathbb{Z}_2, \quad H_0((C_*(f), \partial_*)) \approx \mathbb{Z}.$$

Now recall that $\pi_1(\mathbb{R}P^2) \approx \mathbb{Z}_2$, since its universal cover is S^2, and consider a representation

$$\pi_1(\mathbb{R}P^2, r) \times \mathbb{Z} \to \mathbb{Z}.$$

Denote the gradient flow lines in the above diagram from r to q by $\gamma_{21,r}$ and $\gamma_{21,l}$, from r to p by $\gamma_{20,t}$ and $\gamma_{20,b}$, and the flow lines from q to p by $\gamma_{10,+}$ and $\gamma_{10,-}$. Denote the flow lines in the opposite direction by reversing the subscripts. Choose the constant path at r and the homotopy classes of paths rel endpoints of $\gamma_{21,r}$ and $\gamma_{20,t}$ for the bundle of abelian groups G over $\mathbb{R}P^2$ corresponding to the representation. Note that with this notation the concatenation $\gamma_{21,r}\gamma_{12,l}$ is a generator for $\pi_1(\mathbb{R}P^2, r)$.

If $(1, 1) \mapsto 1$ under the above action, then the homomorphism associated to any path is the identity. Hence, the bundle of abelian groups G over $\mathbb{R}P^2$ determined by the representation is the constant bundle $\mathbb{Z}$, and the Morse-Smale-Witten chain complex with coefficients in G reduces to the above (untwisted) Morse-Smale-Witten chain complex. So, assuming that the action sends $(1, 1) \mapsto -1$, the relevant homomorphisms in the associated bundle of abelian groups are as follows.

$$(\gamma_{12,r})_*(1) = +1$$
$$(\gamma_{12,l})_*(1) = -1$$
$$(\gamma_{01,+})_*(1) = +1$$
$$(\gamma_{01,-})_*(1) = -1$$

Therefore,

$$H_2((C_*(f;G), \partial_*^G)) \approx \mathbb{Z}, \quad H_1((C_*(f;G), \partial_*^G)) \approx 0, \quad H_0((C_*(f;G), \partial_*^G)) \approx \mathbb{Z}_2.$$

The reader can verify that different choices for the homotopy classes of paths rel endpoints from r to q and from r to p yield the same homology. Hence, the homology of $(C_*(f;G), \partial_*^G)$ only depends on the isomorphism class of the bundle of abelian groups over $\mathbb{R}P^2$ when $G_q = \mathbb{Z}$, rather than the specific bundle of abelian groups G.

Now let $\eta \in \Omega_{cl}^1(\mathbb{R}P^2, \mathbb{R})$, and consider the η-twisted Morse-Smale-Witten chain complex $(C_*(f) \otimes \mathbb{R}, \partial_*^\eta)$. For the η-twisted Morse-Smale-Witten boundary operator we have the following.

$$\partial_2^\eta(r) = \exp\left(\int_\mathbb{R} \gamma_{21,r}^*(\eta)\right) q + \exp\left(\int_\mathbb{R} \gamma_{21,l}^*(\eta)\right) q$$

$$\partial_1^\eta(q) = \exp\left(\int_\mathbb{R} \gamma_{10,+}^*(\eta)\right) p - \exp\left(\int_\mathbb{R} \gamma_{10,-}^*(\eta)\right) p$$

Hence, $\partial_2^\eta : C_2(f) \otimes \mathbb{R} \to C_1(f) \otimes \mathbb{R}$ is surjective for any $\eta \in \Omega_{cl}^1(\mathbb{R}P^2, \mathbb{R})$. If η is exact, then it is clear that $\partial_1^\eta = 0$. If η were not exact, then $\partial_1^\eta : C_1(f) \otimes \mathbb{R} \to C_0(f) \otimes \mathbb{R}$ would be surjective. However, $H^1(\mathbb{R}P^2; \mathbb{R}) = 0$, and hence $\partial_1^\eta = 0$ for all $\eta \in \Omega_{cl}^1(\mathbb{R}P^2, \mathbb{R})$. Thus for any $\eta \in \Omega_{cl}^1(\mathbb{R}P^2, \mathbb{R})$,

$$H_2((C_*(f) \otimes \mathbb{R}, \partial_*^\eta)) \approx 0, \quad H_1((C_*(f) \otimes \mathbb{R}, \partial_*^\eta)) \approx 0, \quad H_0((C_*(f) \otimes \mathbb{R}, \partial_*^\eta)) \approx \mathbb{R}.$$

2.5 Computation of $H_0((C_*(f;G), \partial_*^G))$

Let G be a bundle of abelian groups over a closed smooth Riemannian manifold M of dimension $m < \infty$, and let $f : M \to \mathbb{R}$ be a smooth Morse-Smale function on M. Choose a basepoint $x_0 \in Cr_0(f)$ of M. Recall that if M is connected then the isomorphism class of G is determined by a representation

$$\pi_1(M, x_0) \times G_{x_0} \to G_{x_0},$$

and let $H_{x_0} \subseteq G_{x_0}$ be the subgroup generated by elements of the form $g - \gamma_*(g)$ where $g \in G_{x_0}$ and $[\gamma] \in \pi_1(M, x_0)$. The following theorem gives the 0-dimensional twisted Morse homology group of M in terms of the above action, cf. Theorem VI.3.2 of [93].

Theorem 2.16 *If M is connected, then the 0-dimensional twisted Morse homology group of M is isomorphic to G_{x_0}/H_{x_0}, i.e.*

$$H_0((C_*(f;G), \partial_*^G)) \approx G_{x_0}/H_{x_0}.$$

Proof Let $p_0 \in Cr_0(f)$ and pick any path from p_0 to x_0. For all critical points $p_k \in Cr_k(f)$ with $k \geq 2$ the path is homotopic rel endpoints to a path that does not intersect $W^s(p_k)$, since $\dim W^s(p_k) = m - k \leq m - 2$, cf. Theorems 5.16 and 5.17 of [8]. The flow of $-\nabla f$ then gives a homotopy rel endpoints to a path from p_0 to x_0 that lies within an ε-neighborhood of the 1-skeleton of f, i.e.

$$\bigcup_{q \in Cr_1(f)} \overline{W^u(q)}$$

for any $\varepsilon > 0$. An additional homotopy rel endpoints then produces a path γ from p_0 to x_0 that lies in the 1-skeleton of f. Thus, every critical point $p_0 \in Cr_0(f)$ is connected by a path γ to $x_0 \in Cr_0(f)$ that lies in the 1-skeleton of f.

Now, let $Z_0(f;G) = C_0(f;G) = \ker \partial_0^G$ denote the group of 0-cycles and $B_0(f;G) = \mathrm{Im}\, \partial_1^G$ the group of 0-boundaries. We claim that every element of $Z_0(f;G)/B_0(f;G)$ can be represented by an elementary cycle supported at x_0, i.e. gx_0 for some $g \in G_{x_0}$. To see this, let $g_0 p_0$ be an elementary cycle and consider a path γ contained in the 1-skeleton of f connecting p_0 to x_0. Then

$$\mathrm{Im}\, \gamma = \bigcup_{j=1}^{n} \overline{W^u(q_j)}$$

for some critical points $q_j \in Cr_1(f)$. Number the critical points consecutively along Im γ starting with q_1 as the critical point with $p_0 \in \overline{W^u(q_1)}$. Let v be the gradient flow line from q_1 to p_0, and let $p_1 \in \overline{W^u(q_1)}$ be the next element of $Cr_0(f)$ along the path γ connecting p_0 to x_0. Then

$$g_0 p_0 - \partial_1^G \left(\epsilon(v)(\gamma_v)_*(g_0) q_1 \right)$$

has support at p_1, where γ_v is any continuous path from q_1 to p_0 whose image coincides with the image of v. By consecutively subtracting boundary elements of this type for $q_2, \ldots, q_n$ we get a cycle in the same equivalence class of $[g_0 p_0] \in Z_0(f; G)/B_0(f; G)$ with support at x_0. Applying this algorithm to each term in the sum defining an arbitrary element of $Z_0(f; G)$ establishes the claim.

To finish the proof of the theorem, note that any boundary that is supported at x_0 must be the image under ∂_1^G of a chain $\sum_{j=1}^n g_j q_j \in C_1(f; G)$, where

$$\bigcup_{j=1}^n \overline{W^u(q_j)}$$

is the union of the image of a finite number of loops $\gamma_1, \ldots, \gamma_i$ based at x_0. Moreover, since the coefficients in front of the index zero critical points not equal to x_0 in the sum for $\partial_1^G \left(\sum_{j=1}^n g_j q_j \right)$ are all zero, we must have

$$\partial_1^G \left(\sum_{j=1}^n g_j q_j \right) = (g_1 - (\gamma_1)_*(g_1)) x_0 + \cdots + (g_i - (\gamma_i)_*(g_i)) x_0$$

for some $g_1, \ldots, g_i \in G_{x_0}$. □

Remark 2.17 The preceding theorem shows that $H_0((C_*(f; G), \partial_*^G))$ is independent of the Morse-Smale pair (f, g) and depends only on the representation $\pi_1(M, x_0) \to \mathrm{Aut}(G_{x_0})$, i.e. the isomorphism class of G. Also, note that the fact that every path from $x_0 \in Cr_0(f)$ to $p_0 \in Cr_0(f)$ is homotopic rel endpoints to a path that lies in the 1-skeleton of f follows from the Cellular Approximation Theorem, cf. Theorem IV.11.4 of [17], and the results about unstable manifolds and CW-structures discussed in Sect. 4.3.

2.6 The Morse Eilenberg Theorem

In this section we prove a Morse theoretic version of Eilenberg's theorem relating the homology with local coefficients of a space to its equivariant homology. Before we state the theorem we first discuss the Morse-Smale-Witten chain complex on a covering space $\widetilde{M}$ of a closed smooth Riemannian manifold M when the coordinate

2.6.1 The Morse-Smale-Witten Chain Complex on a Covering Space

Let (M, g) be a closed smooth Riemannian manifold of dimension $m < \infty$, and let $f : M \to \mathbb{R}$ be a smooth Morse-Smale function on M. Let $\pi : \widetilde{M} \to M$ be a covering space of M with the coordinate charts on M pulled back to $\widetilde{M}$ so that π is a local diffeomorphism. Let $\widetilde{\mathsf{g}}$ be the pullback of g to $\widetilde{M}$, and let $\tilde{f} = f \circ \pi$.

$$\tilde{f} : (\widetilde{M}, \widetilde{\mathsf{g}}) \xrightarrow{\pi} (M, \mathsf{g}) \xrightarrow{f} \mathbb{R}$$

Nondegeneracy and Morse-Smale transversality are local conditions, and hence $\tilde{f}$ is a smooth Morse-Smale function on $(\widetilde{M}, \widetilde{\mathsf{g}})$. Moreover, for all $k = 0, \ldots, m$ the set of critical points of $\tilde{f}$ of index k is

$$Cr_k(\tilde{f}) = \{\tilde{q} \in \widetilde{M} | \, \pi(\tilde{q}) = q \text{ where } q \in Cr_k(f)\},$$

because the Hessian of $\tilde{f}$ at $\tilde{q}$ is the pullback of the Hessian of f at $q = \pi(\tilde{q})$.

If $\widetilde{M}$ is compact, then the Morse-Smale-Witten chain complex $(C_*(\tilde{f}), \partial_*)$ of $\tilde{f}$ is well-defined, and its homology is isomorphic to the singular homology $H_*(\widetilde{M}; \mathbb{Z})$ by the Morse Homology Theorem, cf. Theorem 7.4 of [8].

Example 2.18 (The Universal Cover of a Real Projective Space) Consider the Morse-Smale-Witten chain complex of the function $f : \mathbb{R}P^2 \to \mathbb{R}$ described in Example 2.15. The function f lifts to a function $\tilde{f} : S^2 \to \mathbb{R}$ on the universal cover S^2 of $\mathbb{R}P^2$, and we can pull back the orientations of the local unstable manifolds of f to those of $\tilde{f} : S^2 \to \mathbb{R}$. The resulting phase diagram of $\tilde{f}$ and the signs associated to the gradient flow lines between critical points of relative index one are shown in the diagram.

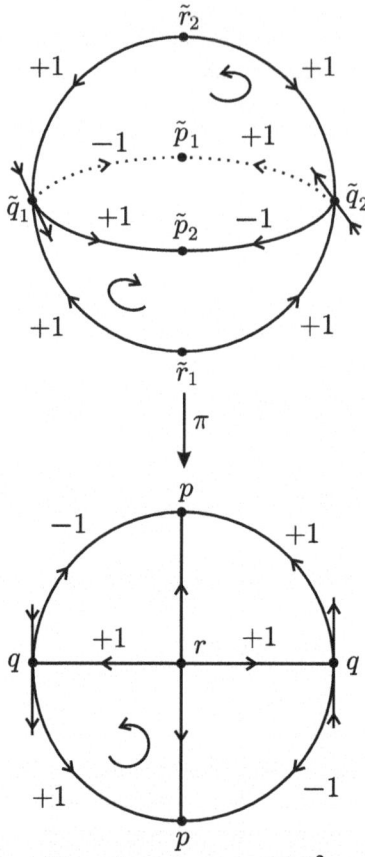

The universal cover of $\mathbb{R}P^2$

We have $Cr_2(\tilde{f}) = \{\tilde{r}_1, \tilde{r}_2\}$ with

$$\tilde{\partial}_2(\tilde{r}_1) = \tilde{\partial}_2(\tilde{r}_2) = \tilde{q}_1 + \tilde{q}_2,$$

and $Cr_1(\tilde{f}) = \{\tilde{q}_1, \tilde{q}_2\}$ with

$$\tilde{\partial}_1(\tilde{q}_1) = \tilde{p}_2 - \tilde{p}_1 \quad \text{and} \quad \tilde{\partial}_1(\tilde{q}_2) = \tilde{p}_1 - \tilde{p}_2,$$

where $Cr_0(\tilde{f}) = \{\tilde{p}_1, \tilde{p}_2\}$. Therefore,

$$H_2((C_*(\tilde{f}), \tilde{\partial}_*)) = <\tilde{r}_1 - \tilde{r}_2> \approx \mathbb{Z}$$
$$H_1((C_*(\tilde{f}), \tilde{\partial}_*)) = <\tilde{q}_1 + \tilde{q}_2> / <\tilde{q}_1 + \tilde{q}_2> \approx 0$$
$$H_0((C_*(\tilde{f}), \tilde{\partial}_*)) = <\tilde{p}_2, \tilde{p}_1> / <\tilde{p}_1 - \tilde{p}_2> \approx \mathbb{Z}$$

as expected.

2.6 The Morse Eilenberg Theorem

We will now show that even if the covering space $\widetilde{M}$ is not compact, then $\tilde{\partial}_*$ is still a well-defined boundary operator and $H_k((C_*(\tilde{f}), \tilde{\partial}_*)) \approx H_k(\widetilde{M}; \mathbb{Z})$ for all $k = 0, \ldots, m$. To see this, first recall that for all $k = 0, \ldots, m$, the group $C_k(\tilde{f})$ is the free abelian group generated by the critical points of index k. Hence, $C_k(\tilde{f})$ consists of sums of the form

$$\sum_{\tilde{q} \in Cr_k(\tilde{f})} n_{\tilde{q}} \tilde{q},$$

where all but finitely many of the coefficients $n_{\tilde{q}} \in \mathbb{Z}$ are zero. So, even if $Cr_k(\tilde{f})$ is infinite, $C_k(\tilde{f})$ consists of finite sums. However, in order to know that $\tilde{\partial}_*$ is well-defined we need to know that if $\tilde{q} \in Cr_k(\tilde{f})$ and $\tilde{p} \in Cr_{k-1}(\tilde{f})$, then the number of gradient flow lines from $\tilde{q}$ to $\tilde{p}$ is finite, for all $1 \leq k \leq m$. To see this, note that a gradient flow line of $(\tilde{f}, \tilde{g})$ from $\tilde{q}$ to $\tilde{p}$ projects to a gradient flow line of (f, g) from $q = \pi(\tilde{q})$ to $p = \pi(\tilde{p})$. Moreover, every gradient flow line of (f, g) from q to p lifts to a unique gradient flow line of $(\tilde{f}, \tilde{g})$ starting at $\tilde{q}$ by the path lifting property of $\pi: \widetilde{M} \to M$. Therefore, the number of gradient flow lines from $\tilde{q}$ to $\tilde{p}$ is less than or equal to the number of gradient flow lines from q to p, which is finite since M is compact and (f, g) satisfies the Morse-Smale transversality condition, cf. Corollary 6.29 of [8].

Example 2.19 (The Universal Cover of a Circle) Consider the height function on S^1 with its universal cover $\mathbb{R}$, where the arrows in the following diagram indicate the direction of minus the gradient flow.

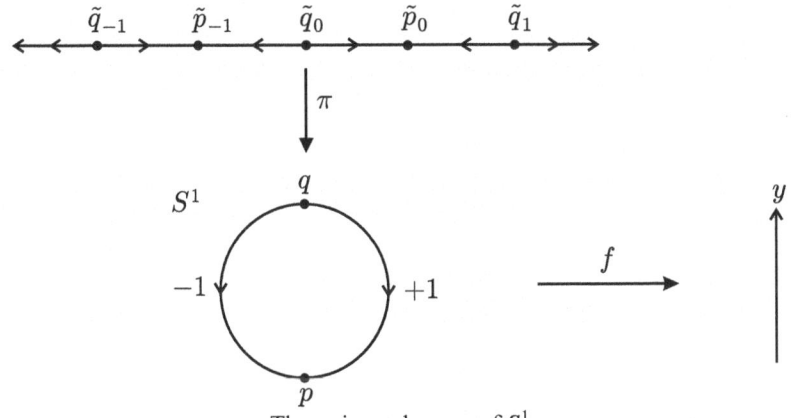

The universal cover of S^1

Above the index 1 critical point q of f there are an infinite number of index 1 critical points $\{\ldots, \tilde{q}_{-1}, \tilde{q}_0, \tilde{q}_1, \ldots\}$ of $\tilde{f} = f \circ \pi$. Similarly, there are an infinite number of index 0 critical points $\{\ldots, \tilde{p}_{-1}, \tilde{p}_0, \tilde{p}_1, \ldots\}$ above the index 0 critical point p. Note that there are two gradient flow lines from q to p, but when these gradient flow lines are lifted to the universal cover starting at a critical point $\tilde{q}_j$ the gradient flow lines in the universal cover end at different critical points $\tilde{p}_{j-1}$ and $\tilde{p}_j$. So, for any $j, j' \in \mathbb{Z}$ the number of gradient flow lines from $\tilde{q}_j$ to $\tilde{p}_{j'}$ is strictly less than the number of gradient flow lines from q to p.

If we pull back the orientations of $W^u(q)$ and $W^u(p)$ to the universal cover, then the signs associated to the gradient flow lines in the universal cover are the same as the signs associated to the gradient flow lines that they correspond to under π. Thus, for any $j \in \mathbb{Z}$ we have

$$\tilde{\partial}_1(\tilde{q}_j) = \tilde{p}_j - \tilde{p}_{j-1},$$

which implies that $H_0((C_*(\tilde{f}), \tilde{\partial}_*)) \approx \mathbb{Z}$. It also implies that for any element in $C_1(\tilde{f})$ we have

$$\tilde{\partial}_1 \left(\sum_{j \in \mathbb{Z}} n_j \tilde{q}_j \right) \neq 0,$$

since only finitely many of the n_j are nonzero. That is, there is a largest j such that $n_j \neq 0$, and thus $n_j \tilde{p}_j \neq 0$ in the sum $\sum_{j \in \mathbb{Z}} n_j \tilde{\partial}_1(\tilde{q}_j)$. This implies that $H_1((C_*(\tilde{f}), \tilde{\partial}_*)) \approx 0$.

Theorem 2.20 (Morse Homology Theorem for a Covering Space) *Let $\pi : \tilde{M} \to M$ be a covering space of a closed smooth Riemannian manifold (M, g) of finite dimension m. Assume that the smooth structure on $\tilde{M}$ is given by pulling back the coordinate charts on the base M, the metric $\tilde{g}$ on $\tilde{M}$ is the pullback of the metric g on M, and $\tilde{f} : \tilde{M} \to \mathbb{R}$ is the pullback of a function $f : M \to \mathbb{R}$. If (f, g) is a Morse-Smale pair on M, then $(\tilde{f}, \tilde{g})$ is a Morse-Smale pair on $(\tilde{M}, \tilde{g})$, the Morse-Smale-Witten chain complex $(C_*(\tilde{f}), \tilde{\partial}_*)$ is well-defined, and its homology is isomorphic to the singular homology $H_*(\tilde{M}; \mathbb{Z})$.*

Proof As discussed above, the pair $(\tilde{f}, \tilde{g})$ is Morse-Smale since π is a local isometry, and $\tilde{\partial}_*$ is well-defined because of the path lifting property. The proof that $H_k((C_*(\tilde{f}), \tilde{\partial}_*)) \approx H_k(\tilde{M}; \mathbb{Z})$ for all $k = 0, \ldots, m$ is essentially the same as the proof of The Morse Homology Theorem for finite dimensional closed smooth manifolds given in Chapter 7 of [8] following [73].

More explicitly, a filtration of M

$$\emptyset = N_{-1} \subseteq N_0 \subseteq N_1 \subseteq \cdots \subseteq N_m = M$$

such that (N_k, N_{j-1}) is a cofibered index pair for

$$W(k, j) = \bigcup_{j \leq \lambda_p \leq \lambda_q \leq k} W(q, p) \subseteq M$$

2.6 The Morse Eilenberg Theorem

for all $0 \leq j \leq k \leq m$, where $W(q, p) = W^u(q) \cap W^s(q) \subset M$, pulls back under $\pi : \widetilde{M} \to M$ to a filtration of $\widetilde{M}$

$$\emptyset = \widetilde{N}_{-1} \subseteq \widetilde{N}_0 \subseteq \widetilde{N}_1 \subseteq \cdots \subseteq \widetilde{N}_m = \widetilde{M}$$

such that $(\widetilde{N}_k, \widetilde{N}_{j-1})$ is a cofibered index pair for

$$\widetilde{W}(k, j) = \bigcup_{j \leq \lambda_{\tilde{p}} \leq \lambda_{\tilde{q}} \leq k} \widetilde{W}(\tilde{q}, \tilde{p}) \subseteq \widetilde{M}$$

for all $0 \leq j \leq k \leq m$, where $\widetilde{W}(\tilde{q}, \tilde{p}) = W^u(\tilde{q}) \cap W^s(\tilde{p}) \subset \widetilde{M}$. One can then show that the following diagram commutes for all $k = 1, \ldots, m$,

$$\begin{array}{ccc} C_k(\tilde{f}) & \xrightarrow{\tilde{\partial}_k} & C_{k-1}(\tilde{f}) \\ \uparrow \approx & & \uparrow \approx \\ H_k(\widetilde{N}_k, \widetilde{N}_{k-1}) & \xrightarrow{\tilde{\delta}_*} & H_{k-1}(\widetilde{N}_{k-1}, \widetilde{N}_{k-2}) \end{array}$$

where $\tilde{\delta}_*$ is the connecting homomorphism of the triple $(\widetilde{N}_k, \widetilde{N}_{k-1}, \widetilde{N}_{k-2})$. The proof is essentially the same as the argument used to prove Lemma 7.21 of [8], which is local in the sense that it works on an isolating neighborhood of $\widetilde{W}(\tilde{q}, \tilde{p}) \cup \{\tilde{q}, \tilde{p}\}$, where $\tilde{q} \in C_k(\tilde{f})$ and $\tilde{p} \in C_{k-1}(\tilde{f})$. Also, note that although [8] assumes that M is orientable, which determines an orientation on the stable manifolds, these orientations are not used in the proof of Lemma 7.21 since it is the orientations of the unstable manifolds that determine generators for $H_k(\widetilde{N}_k, \widetilde{N}_{k-1})$ and $H_{k-1}(\widetilde{N}_{k-1}, \widetilde{N}_{k-2})$. The rest of the proof is identical to the proof of Theorem 7.4 of [8]. □

Note A similar result is stated in Section 6.4.3 of [61], where A. Pajitnov refers to Morse complex on the universal cover as the "universal Morse complex." Pajitnov also proved that the universal Morse complex is simply homotopy equivalent to the chain complex of a triangulation, cf. Theorem A.5 of [64].

2.6.2 The Morse Eilenberg Theorem

Let $f : M \to \mathbb{R}$ be a smooth Morse function on a finite dimensional closed connected smooth manifold M. Pick a basepoint $x_0 \in Cr_0(f)$ for M and a basepoint $\tilde{x}_0 \in \pi^{-1}(x_0)$ for the universal cover $\pi : \widetilde{M} \to M$. Assume that the smooth structure on $\widetilde{M}$ is given by pulling back the charts on M so that π is a local diffeomorphism, and let $\tilde{f} = f \circ \pi$.

The action of $\pi_1(M, x_0)$ on the universal cover by deck transformations

$$\pi_1(M, x_0) \times \tilde{M} \to \tilde{M}$$

restricts to an action on $Cr_k(\tilde{f})$ because

$$Cr_k(\tilde{f}) = \bigcup_{q \in Cr_k(f)} \pi^{-1}(q),$$

and hence there is left action

$$\pi_1(M, x_0) \times C_k(\tilde{f}) \to C_k(\tilde{f})$$

on the free abelian group generated by the critical points of $\tilde{f}$ of index k. If G is a bundle of abelian groups over M, then there is also a left action

$$\pi_1(M, x_0) \times G_{x_0} \to G_{x_0}$$

given by $[\gamma] \cdot g = \gamma_*(g)$ for all $[\gamma] \in \pi_1(M, x_0)$ and $g \in G_{x_0}$, which is converted to a right action by

$$g \cdot [\gamma] \stackrel{\text{def}}{=} [\gamma]^{-1} \cdot g = \gamma_*^{-1}(g).$$

Now consider $G_{x_0} \otimes_{\mathbb{Z}} C_k(\tilde{f})$, and let $K_k(\tilde{f}; G_{x_0})$ be subgroup generated by elements of the form

$$g \cdot [\gamma] \otimes \tilde{q} - g \otimes [\gamma] \cdot \tilde{q},$$

so that

$$G_{x_0} \otimes_{\pi_1} C_k(\tilde{f}) = \left(G_{x_0} \otimes_{\mathbb{Z}} C_k(\tilde{f})\right) / K_k(\tilde{f}; G_{x_0}),$$

where the first tensor product is over $\pi_1(M, x_0)$. Pick a metric g on M such that (f, g) is a Morse-Smale pair and let $\tilde{\mathsf{g}} = \pi^*(\mathsf{g})$. If we orient the unstable manifolds of $(\tilde{f}, \tilde{\mathsf{g}})$ by pulling back the orientations of the unstable manifolds of (f, g), then the (untwisted) Morse-Smale-Witten boundary operator $\partial_k : C_k(f) \to C_{k-1}(f)$ induces a boundary operator on $C_*(\tilde{f})$. Hence, it induces a boundary operator

$$\tilde{\partial}_k : G_{x_0} \otimes_{\mathbb{Z}} C_k(\tilde{f}) \to G_{x_0} \otimes_{\mathbb{Z}} C_{k-1}(\tilde{f}).$$

This boundary operator commutes with the action of $\pi_1(M, x_0)$ on $\tilde{M}$, and thus it maps $K_k(\tilde{f}; G_{x_0})$ to $K_{k-1}(\tilde{f}; G_{x_0})$. Hence, there is an induced boundary operator

$$\bar{\partial}_k : G_{x_0} \otimes_{\pi_1} C_k(\tilde{f}) \to G_{x_0} \otimes_{\pi_1} C_{k-1}(\tilde{f}).$$

2.6 The Morse Eilenberg Theorem

Theorem 2.21 (Morse Eilenberg) *Let G be a bundle of abelian groups over a closed connected smooth Riemannian manifold $(M, \mathfrak{g})$ of dimension $m < \infty$, and let $f : M \to \mathbb{R}$ be a smooth Morse-Smale function on M. Then there is an isomorphism of chain complexes*

$$(G_{x_0} \otimes_{\pi_1} C_*(\tilde{f}), \bar{\partial}_*) \approx (C_*(f; G), \partial_*^G),$$

and hence

$$H_k((G_{x_0} \otimes_{\pi_1} C_*(\tilde{f}), \bar{\partial}_*)) \approx H_k((C_*(f; G), \partial_*^G))$$

for all $k = 0, \ldots, m$.

Proof Since $\tilde{M}$ is simply connected, for any $\tilde{q} \in Cr_k(\tilde{f})$ there is a unique homotopy class of paths rel endpoints from $\tilde{q}$ to the basepoint $\tilde{x}_0 \in \tilde{M}$, where $0 \leq k \leq m$. So, if $\tilde{\gamma}_{\tilde{q}}$ is any path in $\tilde{M}$ from $\tilde{q}$ to $\tilde{x}_0$, then $\gamma_{\tilde{q}} \equiv \pi \circ \tilde{\gamma}_{\tilde{q}}$ determines a well-defined homotopy class of paths rel endpoints in M from $q \equiv \pi(\tilde{q}) \in Cr_k(f)$ to the basepoint $x_0 = \pi(\tilde{x}_0)$ of M. Thus, there is a homomorphism

$$\tilde{\Phi}_k : G_{x_0} \otimes_\mathbb{Z} C_k(\tilde{f}) \to C_k(f; G)$$

defined on a generator $g \otimes \tilde{q}$ by

$$\tilde{\Phi}_k(g \otimes \tilde{q}) \stackrel{\text{def}}{=} (\gamma_{\tilde{q}})_*(g) q,$$

where $g \in G_{x_0}$ and $\tilde{q} \in Cr_k(\tilde{f})$.

We claim that $K_k(\tilde{f}; G_{x_0})$ is the kernel of $\tilde{\Phi}_k$. To see this, take any generator

$$g \cdot [\gamma] \otimes \tilde{q} - g \otimes [\gamma] \cdot \tilde{q} \in K_k(\tilde{f}; G_{x_0}),$$

where $g \in G_{x_0}$, $\tilde{q} \in Cr_k(\tilde{f})$, and $[\gamma] \in \pi_1(M, x_0)$, and note that $[\gamma] \cdot \tilde{\gamma}_{\tilde{q}}$ is a path from $[\gamma] \cdot \tilde{q}$ to $[\gamma] \cdot \tilde{x}_0$. Lifting the path γ to a path starting at $\tilde{x}_0$ gives a path from $\tilde{x}_0$ to $[\gamma] \cdot \tilde{x}_0$ whose inverse is a path $\tilde{\gamma}_{[\gamma] \cdot \tilde{x}_0}$ from $[\gamma] \cdot \tilde{x}_0$ to $\tilde{x}_0$. The concatenation $([\gamma] \cdot \tilde{\gamma}_{\tilde{q}}) \tilde{\gamma}_{[\gamma] \cdot \tilde{x}_0}$ is a path from $[\gamma] \cdot \tilde{q}$ to $\tilde{x}_0$. Therefore,

$$\tilde{\Phi}_k(g \cdot [\gamma] \otimes \tilde{q} - g \otimes [\gamma] \cdot \tilde{q}) = \tilde{\Phi}_k(g \cdot [\gamma] \otimes \tilde{q}) - \tilde{\Phi}(g \otimes [\gamma] \cdot \tilde{q})$$

$$= (\gamma_{\tilde{q}})_* \left(\gamma_*^{-1}(g) \right) q - \left((\gamma_{\tilde{q}})_* \circ \gamma_*^{-1} \right) (g) q$$

$$= 0,$$

since $\pi \circ ([\gamma] \cdot \tilde{\gamma}_{\tilde{q}}) = \gamma_{\tilde{q}}$ and $\pi \circ \tilde{\gamma}_{[\gamma] \cdot \tilde{x}_0} = \gamma^{-1}$.

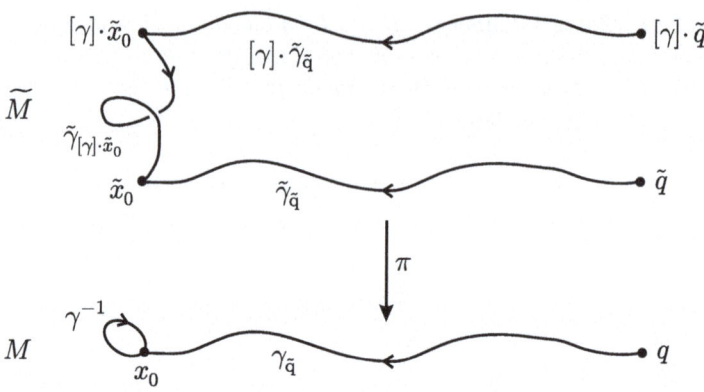

Moreover, if $\tilde{\Phi}_k$ sends some element of $G_{x_0} \otimes_\mathbb{Z} C_k(\tilde{f})$ to $0 \in C_k(f; G)$, then the coefficients in front of each element of $Cr_k(f)$ in the image must be zero. Thus, we may assume that the element that $\tilde{\Phi}_k$ maps to zero is of the form $\sum_{i=1}^n g_i \otimes \tilde{q}_i$, where $\pi(\tilde{q}_i) = q \in Cr_k(f)$ for all $i = 1, \ldots, n$. Then for all $i = 2, \ldots, n$, there exist $[\gamma_i] \in \pi_1(M, x_0)$ such that $\tilde{q}_i = [\gamma_i] \cdot \tilde{q}_1$. Hence,

$$0 = \tilde{\Phi}_k \left(g_1 \otimes \tilde{q}_1 + \sum_{i=2}^n g_i \otimes [\gamma_i] \cdot \tilde{q}_1 \right)$$
$$= (\gamma_{\tilde{q}_1})_*(g_1) q + \sum_{i=2}^n \left((\gamma_{\tilde{q}_1})_* \circ (\gamma_i^{-1})_* \right) (g_i) q$$
$$= (\gamma_{\tilde{q}_1})_* \left(g_1 + \sum_{i=2}^n (\gamma_i^{-1})_*(g_i) \right) q,$$

which implies that

$$g_1 + \sum_{i=2}^n (\gamma_i^{-1})_*(g_i) = 0,$$

since $(\gamma_{\tilde{q}_1})_*$ is an isomorphism. Therefore,

$$\sum_{i=1}^n g_i \otimes \tilde{q}_i = g_1 \otimes \tilde{q}_1 + \sum_{i=2}^n g_i \otimes [\gamma_i] \cdot \tilde{q}_1$$
$$= -\sum_{i=2}^n (\gamma_i^{-1})_*(g_i) \otimes \tilde{q}_1 + \sum_{i=2}^n g_i \otimes [\gamma_i] \cdot \tilde{q}_1$$

2.6 The Morse Eilenberg Theorem

$$= \sum_{i=2}^{n} \left(g_i \otimes [\gamma_i] \cdot \tilde{q}_1 - g_i \cdot [\gamma_i] \otimes \tilde{q}_1 \right)$$

$$\in K_k(\tilde{f}; G_{x_0}).$$

We have shown that $K_k(\tilde{f}; G_{x_0}) = \ker \tilde{\Phi}_k$. To see that $\tilde{\Phi}_k$ is surjective note that for any elementary chain $gq \in C_k(f; G)$ we can pick a path γ_q^{-1} from x_0 to q and lift it to a path in $\tilde{M}$ starting at $\tilde{x}_0$. The end of the lifted path will be a critical point $\tilde{q}$ with $\pi(\tilde{q}) = q$, and $\tilde{\Phi}_k \left((\gamma_q^{-1})_*(g) \otimes \tilde{q} \right) = (\gamma_{\tilde{q}})_*((\gamma_q^{-1})_*(g))q = gq$. Therefore, we have an induced isomorphism

$$\Phi_k : \left(G_{x_0} \otimes_{\mathbb{Z}} C_k(\tilde{f}) \right) / K_k(\tilde{f}; G_{x_0}) \to C_k(f; G)$$

for all $k = 0, \ldots, m$.

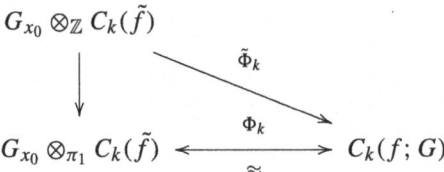

It remains to show that Φ_k is a chain map for all $0 \leq k \leq m$. To do this, we will show that $\tilde{\Phi}_k$ is a chain map. It then follows that Φ_k is a chain map because the boundary operator $\bar{\partial}_*$ on $G_{x_0} \otimes_{\pi_1} C_*(\tilde{f})$ is induced from the boundary operator ∂_* on $G_{x_0} \otimes_{\mathbb{Z}} C_*(\tilde{f})$. So, let $g \otimes \tilde{q}$ be a generator of $G_{x_0} \otimes_{\mathbb{Z}} C_k(\tilde{f})$, and note that

$$\partial_k^G(\tilde{\Phi}_k(g \otimes \tilde{q})) = \partial_k^G((\gamma_{\tilde{q}})_*(g)q)$$
$$= \sum_{p \in Cr_{k-1}(f)} \sum_{v \in \mathcal{M}(q,p)} \epsilon(v) \gamma_*^v((\gamma_{\tilde{q}})_*(g))p$$

where $\gamma^v : [0, 1] \to M$ is any continuous path from p to q whose image coincides with the image of $v \in \mathcal{M}(q, p)$ and $\epsilon(v) = \pm 1$ is the sign determined by the orientation on $\mathcal{M}(q, p)$. On the other hand,

$$\tilde{\Phi}_{k-1}\left(\bar{\partial}_k(g \otimes \tilde{q}) \right) = \tilde{\Phi}_{k-1}\left(g \otimes \sum_{\tilde{p} \in Cr_{k-1}(\tilde{f})} \sum_{\tilde{v} \in \mathcal{M}(\tilde{q}, \tilde{p})} \epsilon(\tilde{v}) \tilde{p} \right)$$
$$= \sum_{\tilde{p} \in Cr_{k-1}(\tilde{f})} \sum_{\tilde{v} \in \mathcal{M}(\tilde{q}, \tilde{p})} \epsilon(\tilde{v}) \tilde{\Phi}_{k-1}(g \otimes \tilde{p})$$

$$= \sum_{\tilde{p} \in Cr_{k-1}(\tilde{f})} \sum_{\tilde{\nu} \in \mathcal{M}(\tilde{q},\tilde{p})} \epsilon(\tilde{\nu})(\gamma_{\tilde{p}})_*(g) p$$

$$= \sum_{\tilde{p} \in Cr_{k-1}(\tilde{f})} \sum_{\tilde{\nu} \in \mathcal{M}(\tilde{q},\tilde{p})} \epsilon(\tilde{\nu})(\pi \circ \gamma^{\tilde{\nu}} \tilde{\gamma}_{\tilde{q}})_*(g)) p$$

$$= \sum_{\tilde{p} \in Cr_{k-1}(\tilde{f})} \sum_{\tilde{\nu} \in \mathcal{M}(\tilde{q},\tilde{p})} \epsilon(\tilde{\nu})(\pi \circ \gamma^{\tilde{\nu}})_*((\gamma_{\tilde{q}})_*(g)) p,$$

where $\gamma^{\tilde{\nu}} : [0, 1] \to \widetilde{M}$ is any continuous path from $\tilde{p}$ to $\tilde{q}$ whose image coincides with the image of $\tilde{\nu} \in \mathcal{M}(\tilde{q}, \tilde{p})$, $\epsilon(\tilde{\nu}) = \pm 1$ is the sign determined by the orientation on $\mathcal{M}(\tilde{q}, \tilde{p})$, and $\gamma^{\tilde{\nu}}\tilde{\gamma}_{\tilde{q}}$ (the concatenation of two paths in $\widetilde{M}$) is a path from $\tilde{p}$ to $\tilde{x}_0$.

Finally, note that we can pick the paths $\gamma^{\tilde{\nu}}$ in the sum for $\tilde{\Phi}_{k-1}\left(\tilde{\partial}_k(g \otimes \tilde{q})\right)$ to be lifts of the paths γ^{ν} in the sum for $\partial_k^G(\tilde{\Phi}_k(g \otimes \tilde{q}))$. Moreover, there is a bijection between the collection of paths γ^{ν} in the first sum and the collection of paths $\gamma^{\tilde{\nu}}$ in the second sum with $\epsilon(\nu) = \epsilon(\tilde{\nu})$ if $\gamma^{\tilde{\nu}}$ is a lift of γ^{ν}, since the orientations of the unstable manifolds of $(\tilde{f}, \tilde{g})$ were determined by pulling back the orientations of the unstable manifolds of (f, g). Therefore, $\partial_k^G(\tilde{\Phi}_k(g \otimes \tilde{q})) = \tilde{\Phi}_{k-1}\left(\tilde{\partial}_k(g \otimes \tilde{q})\right)$.

This completes the proof of Theorem 2.21. □

Remark 2.22 (Basepoints and the Isomorphism Class of G) Note that the chain complex $(G_{x_0} \otimes_{\pi_1} C_*(\tilde{f}), \tilde{\partial}_*)$ was defined using a basepoint $x_0 \in M$, whereas the definition of $(C_*(f; G), \partial_*^G)$ does not involve x_0. Thus, the Morse Eilenberg Theorem (Theorem 2.21) implies that the isomorphism class of $H_k((G_{x_0} \otimes_{\pi_1} C_*(\tilde{f}), \tilde{\partial}_*))$ does not depend on the basepoint $x_0 \in M$ for all $k = 0, \ldots, m$.

On the other hand, the boundary operator of the twisted Morse-Smale-Witten chain complex $(C_*(f; G), \partial_*^G)$ was defined using the homomorphisms that define the bundle of abelian groups G (Definition 2.11), whereas the boundary operator for the chain complex $(G_{x_0} \otimes_{\pi_1} C_*(\tilde{f}), \tilde{\partial}_*)$ was defined using the representation $\pi_1(M, x_0) \times G_{x_0} \to G_{x_0}$ induced by the homomorphisms that define G (Theorem 2.3). Thus, the Morse Eilenberg Theorem shows that the homology of the twisted Morse-Smale-Witten chain complex $(C_*(f; G), \partial_*^G)$ only depends on the isomorphism class of G.

We will give an independent proof of this fact in the next chapter, where we also show that the homology does not depend on the Morse-Smale pair (f, g). We prove a similar isomorphism directly for twisted Morse cohomology in Proposition 5.4.

Remark 2.23 (Eilenberg's Theorem and Twisted Morse Homology) Let (X, x_0) be a pointed connected topological space and G_{x_0} an abelian group with a $\pi_1(X, x_0)$ action. The equivariant homology groups $E_*(\widetilde{X}; G_{x_0})$ are defined to be the homology groups of a chain complex $(G_{x_0} \otimes_{\pi_1} C_*(\widetilde{X}), \tilde{\partial}_*)$, where $C_*(\widetilde{X})$ denotes the singular chains in the universal cover $\widetilde{X}$ and the boundary operator $\tilde{\partial}_*$ is induced from the singular boundary operator on $C_*(\widetilde{X})$. The action of $\pi_1(X, x_0)$ on

2.6 The Morse Eilenberg Theorem

G_{x_0} can be used to define a bundle of abelian groups G on X in the isomorphism class determined by the action (Theorem 2.3), and G can then be used to define homology groups $H_*(X; G)$ with coefficients in the bundle G. A well-known theorem due to Eilenberg says that

$$E_k(\widetilde{X}; G_{x_0}) \approx H_k(X; G)$$

for all k, which implies that the homology groups $H_*(X; G)$ only depend on the isomorphism class of G (see Sect. 4.1).

If $(X, x_0) = (M, x_0)$ is a finite dimensional closed connected smooth manifold and $(f, \mathfrak{g})$ is a Morse-Smale pair on M, then the coordinate charts on M and the pair $(f, \mathfrak{g})$ can be pulled back to the universal cover of M using the projection map $\pi : \widetilde{M} \to M$. Under these conditions, the Morse Eilenberg Theorem (Theorem 2.21) says that there is an isomorphism

$$H_k((G_{x_0} \otimes_{\pi_1} C_*(\tilde{f}), \bar{\partial}_*)) \approx H_k((C_*(f; G), \partial_*^G))$$

for all k. One approach to proving the Twisted Morse Homology Theorem (Theorem 4.1) would be to directly establish the existence of an isomorphism

$$H_*((G_{x_0} \otimes_{\pi_1} C_*(\tilde{f}), \bar{\partial}_*)) \xleftarrow{\quad ? \quad} E_*(\widetilde{M}; G_{x_0}),$$

since we would then have the following isomorphisms for all k.

$$\begin{array}{ccc}
H_k((G_{x_0} \otimes_{\pi_1} C_*(\tilde{f}), \bar{\partial}_*)) & \xleftarrow{\approx} & E_k(\widetilde{M}; G_{x_0}) \\
\approx \Big\downarrow \text{Theorem 2.21} & & \approx \Big\uparrow \text{Eilenberg} \\
H_k((C_*(f; G), \partial_*^G)) & \xleftarrow[\text{Theorem 4.1}]{\approx} & H_k(M; G)
\end{array}$$

Theorem 2.20 proves the existence of an isomorphism between the Morse homology and the singular homology of the universal cover using Conley index theory,

$$H_*(C_*(\tilde{f}), \bar{\partial}_*) \xleftarrow{\approx} H_*(\widetilde{M}; \mathbb{Z}),$$

but that result does not address the actions of $\pi_1(M, x_0)$ on $\widetilde{M}$ and G_{x_0}. A. Pajitnov proved a similar isomorphism in the context of Novikov theory using a smooth triangulation of $\widetilde{M}$ (Theorem A.5 of [64]), but the actions of $\pi_1(M, x_0)$ on $\widetilde{M}$ and G_{x_0} were not considered in that result either.

On the other hand, both Theorem 2.21 and Eilenberg's Theorem (Theorem 4.3) concern the action of $\pi_1(M, x_0)$ on $\widetilde{M}$ and G_{x_0}, but Theorem 2.21 gives an isomorphism within the category of Morse chain complexes and Eilenberg's Theorem gives an isomorphism within the category of singular chain complexes. Constructing an isomorphism directly for the top row in the above diagram would involved working with Morse chain complexes, singular chain complexes, and the actions of $\pi_1(M, x_0)$ on $\widetilde{M}$ and G_{x_0} altogether.

It seems likely that the proof of Theorem 2.20 and the proof of Theorem A.5 of [64] could be adapted to account for the actions of $\pi_1(M, x_0)$ on $\widetilde{M}$ and G_{x_0}. However, we choose instead to prove the isomorphism in the bottom row of the above diagram (Theorem 4.1) using standard techniques from Floer theory and a combination of techniques from classical algebraic topology, differential topology, and homotopy theory. Our approach to proving the Twisted Morse Homology Theorem (Theorem 4.1) allows for a direct comparison between the twisted Morse-Smale-Witten boundary operator and Steenrod's twisted cellular boundary operator, something that may be of interest to people working in Floer theory. It also involves the construction of a Morse-Smale pair whose unstable manifolds coincide with a smooth triangulation of the manifold (Theorem 4.12), which is essential for the results in Chap. 5 that use Stokes' Theorem (see the discussion at the beginning of Sect. 5.3), and may be of interest to people working with combinatorial Morse theory.

Chapter 3
The Homology Determined by the Isomorphism Class of G

In this chapter we prove an invariance theorem which shows that on a closed finite dimensional smooth Riemannian manifold the homology of the twisted Morse-Smale-Witten chain complex is independent of the Morse-Smale pair and depends only on the isomorphism class of the bundle of abelian groups G. The proof of the twisted invariance theorem follows standard arguments used to prove invariance for the Morse-Smale-Witten chain complex with integer coefficients. This chapter contains all the details needed to extend the proof from integer coefficients to local coefficients.

3.1 A Chain Map

Let $\rho : \overline{\mathbb{R}} \to [-1, 1]$ be a smooth strictly increasing function with $\rho(-\infty) = -1$, $\rho(+\infty) = 1$, and $\lim_{t \to \pm\infty} \rho'(t) = 0$, and let (f_1, g_1) and (f_2, g_2) be smooth Morse-Smale pairs on M. By adding a constant to f_1 we may assume that $\inf f_1 > \sup f_2$. Let $F_{21} : \mathbb{R} \times M \to \mathbb{R}$ be a smooth function that is strictly decreasing in its first component such that for some large $T \gg 0$ we have

$$F_{21}(t, x) = \begin{cases} f_1(x) - \rho(t) & \text{if } t < -T \\ h_t(x) & \text{if } -T \leq t \leq T \\ f_2(x) - \rho(t) & \text{if } t > T \end{cases}$$

where $h_t(x)$ is an approximation to $\frac{1}{2T}(T - t)(f_1(x) - \rho(t)) + \frac{1}{2T}(T + t)(f_2(x) - \rho(t))$ with $\frac{d}{dt} h_t(x) < 0$ that makes F_{21} smooth. Extend F_{21} to a smooth function on $\overline{\mathbb{R}} \times M$ by defining $F_{21}(-\infty, x) = f_1(x) + 1$ and $F_{21}(+\infty, x) = f_2(x) - 1$. Note that the critical points of F_{21} are all on the boundary of $\overline{\mathbb{R}} \times M$ since $\frac{d}{dt} F_{21}(t, x) < 0$

for all $t \in \mathbb{R}$. Pick a Riemannian metric g on $\overline{\mathbb{R}} \times M$ such that

$$\mathsf{g}(t,x) = \begin{cases} \mathsf{g}_1(x) + dt^2 & \text{if } t < -T \\ \mathsf{g}_2(x) + dt^2 & \text{if } t > T \end{cases}$$

for all $x \in M$. (Compare with Lemma 1.17 of [24].)

Let $\varphi_\tau : \overline{\mathbb{R}} \times M \to \overline{\mathbb{R}} \times M$ denote the flow associated to the negative gradient $-\nabla F_{21}$ with respect to the metric g. Let $q_1 \in Cr(f_1)$, $q_2 \in Cr(f_2)$, and define

$$W_F^u(q_1) = \{(t,x) \in \overline{\mathbb{R}} \times M | \lim_{\tau \to -\infty} \varphi_\tau(t,x) = (-\infty, q_1)\}$$

$$W_F^s(q_2) = \{(t,x) \in \overline{\mathbb{R}} \times M | \lim_{\tau \to +\infty} \varphi_\tau(t,x) = (+\infty, q_2)\}$$

$$W_F(q_1,q_2) = W_F^u(q_1) \cap W_F^s(q_2) \subset \mathbb{R} \times M.$$

Since both (f_1, g_1) and (f_2, g_2) are Morse-Smale pairs we can perturb either

(1) the approximation $h_t(x)$ or (2) the Riemannian metric g

so that $W_F^u(q_1) \pitchfork W_F^s(q_2)$ for all $q_1 \in Cr(f_1)$ and $q_2 \in Cr(f_2)$. (Compare with Proposition 1.12 and Lemma 1.13 of [24].)

Choosing orientations for the unstable manifolds $W^u(q_1)$ and $W^u(q_2)$ then determines an orientation on $W_F(q_1, q_2)$ via the short exact sequence

$$T_*W_F(q_1,q_2) \hookrightarrow T_*W_F^u(q_1)|_{W_F(q_1,q_2)} \to \nu_*(W_F(q_1,q_2), W_F^u(q_1))|_{W_F(q_1,q_2)} \to 0$$

where the fibers of the normal bundle are isomorphic to $T_{q_2} W_{f_2}^u(q_2)$ via the flow of $-\nabla F_{21}$ and $W_F^u(q_1)$ is oriented as follows. For $\tau << 0$, $\varphi_\tau(t,x)$ will be in the region where $t < -T$. In that region the tangent space to $W_F^u(q_1)$ has the product orientation $\mathbb{R} \times T_* W_{f_1}^u(q_1)$, and this orientation then determines an orientation on $T_{(t,x)} W_F^u(q_1)$ via $\varphi_{-\tau}$.

Taking a quotient by the action of $\mathbb{R}$ given by the negative gradient flow then gives an oriented smooth manifold

$$\mathcal{M}_F(q_1, q_2) = (W_F^u(q_1) \cap W_F^s(q_2))/\mathbb{R}$$

of dimension $\lambda_{q_1} - \lambda_{q_2}$, called the moduli space of time dependent gradient flow lines. The orientation on $\mathcal{M}_F(q_1, q_2)$ is chosen by identifying the space with a level set and putting $-\nabla F_{21}$ first, analogous to the way the orientation on $\mathcal{M}(q,p)$ was chosen.

Denote the pullback of a bundle of abelian groups G over M under the projection $\pi : \overline{\mathbb{R}} \times M \to M$ by G^*. Thus, G^* associates to every point $(t,x) \in \overline{\mathbb{R}} \times M$ the abelian group G_x and to every continuous path $\gamma : [0,1] \to \overline{\mathbb{R}} \times M$ the homomorphism $\gamma_* : G_{\pi(\gamma(1))} \to G_{\pi(\gamma(0))}$ determined by the path $\pi \circ \gamma$.

3.1 A Chain Map

Definition 3.1 (Chain Map) Let (f_1, g_1) and (f_2, g_2) be smooth Morse-Smale pairs on M, and let G_1 and G_2 be bundles of abelian groups over M. Assume that G_1 and G_2 are isomorphic, so there exists a family of isomorphisms $\Phi : G_1 \to G_2$ making the diagram in Definition 2.2 commute. Let $(F_{21})_\square : (C_*(f_1; G_1), \partial_*^{G_1}) \to (C_*(f_2; G_2), \partial_*^{G_2})$ be the linear map defined on an elementary chain $gq_1 \in C_k(f_1; G_1)$ by

$$(F_{21})_\square(gq_1) = (-1)^{\lambda_{q_1}} \sum_{q_2 \in Cr_k(f_2)} \sum_{\nu_F \in \mathcal{M}_F(q_1, q_2)} \epsilon(\nu_F)(\gamma_F)_*(\Phi(g))q_2$$

where $\gamma_F : [0, 1] \to \overline{\mathbb{R}} \times M$ is any continuous path from $(+\infty, q_2)$ to $(-\infty, q_1)$ such that the image $\gamma_F((0, 1))$ coincides with $Im(\nu_F)$, $\Phi(g) \in (G_2)_{q_1} = (G_2^*)_{(-\infty, q_1)}$, $\epsilon(\nu_F) = \pm 1$ is the sign determined by the orientation on $\mathcal{M}_F(q_1, q_2)$, and λ_{q_1} denotes the index of q_1 as a critical point of $f_1 : M \to \mathbb{R}$.

Note If $G_1 = G_2 = \mathbb{Z}$ and Φ is the identity, then $(F_{21})_\square(q_1)$ is $(-1)^{\lambda_{q_1}}$ times the continuation map defined in Section 4.1.3 of [78].

The above sum over $\mathcal{M}_F(q_1, q_2)$ is finite because of the following result, cf. Section 2.4.3 of [78].

Theorem 3.2 (Compactification) *For any $q_1 \in Cr(f_1)$ and $q_2 \in Cr(f_2)$ the moduli space $\mathcal{M}_F(q_1, q_2)$ has a compactification $\overline{\mathcal{M}}_F(q_1, q_2)$, consisting of all the piecewise gradient flow lines from $(-\infty, q_1)$ to $(+\infty, q_2)$, including both time dependent and time independent gradient flow lines.*

We will now show how this structure on the 1-dimensional compactified moduli spaces of time dependent gradient flow lines implies that $(F_{21})_\square$ is a chain map. In order to distinguish between the time independent gradient flow lines and the time dependent gradient flow lines, in the following we will denote the (zero dimensional) moduli spaces of time independent gradient flow lines of $f_1 : \{-\infty\} \times M \to \mathbb{R}$ by $\mathcal{M}_{f_1}(q_1, p_1)$, the time independent moduli spaces of gradient flow lines of $f_2 : \{+\infty\} \times M \to \mathbb{R}$ by $\mathcal{M}_{f_2}(q_2, p_2)$, and the time dependent gradient flow lines of $F_{21} : \overline{\mathbb{R}} \times M \to \mathbb{R}$ by $\mathcal{M}_F(q_1, q_2)$ or $\mathcal{M}_F(p_1, p_2)$.

If $\lambda_{q_1} - \lambda_{p_2} = 1$, then $\overline{\mathcal{M}}_F(p_1, p_2)$ is a 1-dimensional smooth manifold with boundary, and

$$\partial \overline{\mathcal{M}}_F(q_1, p_2) = \left(\bigsqcup_{p_1} \mathcal{M}_{f_1}(q_1, p_1) \times \mathcal{M}_F(p_1, p_2) \right)$$
$$\bigsqcup \left(\bigsqcup_{q_2} \mathcal{M}_F(q_1, q_2) \times \mathcal{M}_{f_2}(q_2, p_2) \right).$$

So, there are three possibilities for a nonempty boundary of a path component $\overline{\mathcal{M}}_F(q_1, p_2; [\nu_F])$.

1. We have $\partial \overline{\mathcal{M}}_F(q_1, p_2; [\nu_F]) = \{(\nu_1, \nu_F^p), (\tilde{\nu}_1, \tilde{\nu}_F^p)\}$, where $\nu_1 \in \mathcal{M}_{f_1}(q_1, p_1)$, $\nu_F^p \in \mathcal{M}_F(p_1, p_2)$, $\tilde{\nu}_1 \in \mathcal{M}_{f_1}(q_1, \tilde{p}_1)$, and $\tilde{\nu}_F^p \in \mathcal{M}_F(\tilde{p}_1, p_2)$.

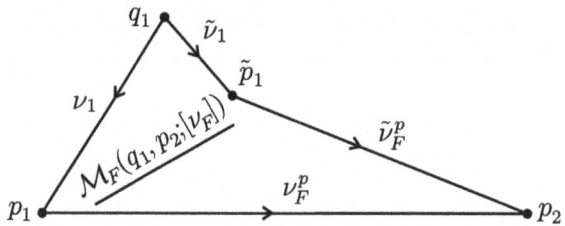

2. We have $\partial \overline{\mathcal{M}}_F(q_1, p_2; [\nu_F]) = \{(\nu_1, \nu_F^p), (\nu_F^q, \nu_2)\}$, where $\nu_1 \in \mathcal{M}_{f_1}(q_1, p_1)$, $\nu_F^p \in \mathcal{M}_F(p_1, p_2)$, $\nu_F^q \in \mathcal{M}_F(q_1, q_2)$, $\nu_2 \in \mathcal{M}_{f_2}(q_2, p_2)$.

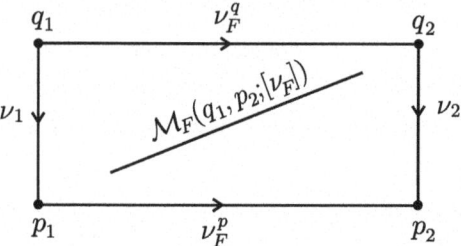

3. We have $\partial \overline{\mathcal{M}}_F(q_1, p_2; [\nu_F]) = \{(\nu_F^q, \nu_2), (\tilde{\nu}_F^q, \tilde{\nu}_2)\}$, where $\nu_F^q \in \mathcal{M}_F(q_1, q_2)$, $\nu_2 \in \mathcal{M}_{f_2}(q_2, p_2)$, $\tilde{\nu}_F^q \in \mathcal{M}_F(q_1, \tilde{q}_2)$, $\tilde{\nu}_2 \in \mathcal{M}_{f_2}(\tilde{q}_2, p_2)$.

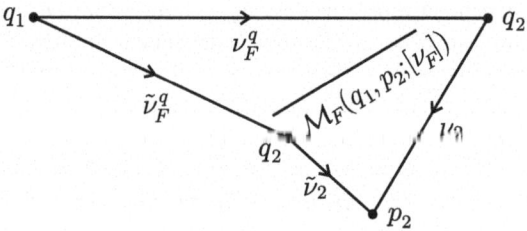

In all three cases we have orientation formulas analogous to those in Lemma 2.9.

1. $\partial \overline{\mathcal{M}}_F(q_1, p_2; [\nu_F]) = (\tilde{\nu}_1, \tilde{\nu}_F^p) - (\nu_1, \nu_F^p)$ and $\epsilon(\tilde{\nu}_1)\epsilon(\tilde{\nu}_F^p) + \epsilon(\nu_1)\epsilon(\nu_F^p) = 0$.
2. $\partial \overline{\mathcal{M}}_F(q_1, p_2; [\nu_F]) = (\nu_F^q, \nu_2) - (\nu_1, \nu_F^p)$ and $\epsilon(\nu_F^q)\epsilon(\nu_2) + \epsilon(\nu_1)\epsilon(\nu_F^p) = 0$.
3. $\partial \overline{\mathcal{M}}_F(q_1, p_2; [\nu_F]) = (\nu_F^q, \nu_2) - (\tilde{\nu}_F^q, \tilde{\nu}_2)$ and $\epsilon(\nu_F^q)\epsilon(\nu_2) + \epsilon(\tilde{\nu}_F^q)\epsilon(\tilde{\nu}_2) = 0$.

These formulas can be verified by first checking them for specific orientations chosen on $W^u(q_1)$, $W^u(q_2)$, $W^u(p_1)$, and $W^u(p_2)$ and then noting that changing an orientation changes a complimentary pair of signs.

Proposition 3.3 *The map* $(F_{21})_\square : (C_*(f_1; G_1), \partial_*^{G_1}) \to (C_*(f_2; G_2), \partial_*^{G_2})$ *is a chain map. That is, the map preserves degree and* $\partial_*^{G_2} \circ (F_{21})_\square = (F_{21})_\square \circ \partial_*^{G_1}$.

3.1 A Chain Map

Proof Let $gq_1 \in C_k(f_1; G_1)$ be an elementary chain for some $k = 1, \ldots, m$ where $m = \dim M$. We have $\partial_k^{G_2}((F_{21})_\square(gq_1))$

$$= \partial_k^{G_2}\left((-1)^{\lambda_{q_1}} \sum_{q_2 \in Cr_k(f_2)} \sum_{v_F^q \in M_F(q_1,q_2)} \epsilon(v_F^q)(\gamma_F^q)_*(\Phi(g))q_2\right)$$

$$= (-1)^k \sum_{\substack{q_2 \in Cr_k(f_2) \\ v_F^q \in M_F(q_1,q_2)}} \epsilon(v_F^q) \partial_k^{G_2}\left((\gamma_F^q)_*(\Phi(g))q_2\right)$$

$$= (-1)^k \sum_{\substack{q_2 \in Cr_k(f_2) \\ v_F^q \in M_F(q_1,q_2)}} \epsilon(v_F^q) \sum_{\substack{p_2 \in Cr_{k-1}(f_2) \\ v_2 \in M_{f_2}(q_2,p_2)}} \epsilon(v_2)(\gamma_2)_*\left((\gamma_F^q)_*(\Phi(g))\right) p_2$$

$$= (-1)^k \sum_{\substack{q_2 \in Cr_k(f_2) \\ v_F^q \in M_F(q_1,q_2)}} \sum_{\substack{p_2 \in Cr_{k-1}(f_2) \\ v_2 \in M_{f_2}(q_2,p_2)}} \epsilon(v_F^q)\epsilon(v_2)(\gamma_2(\pi \circ \gamma_F^q))_*(\Phi(g))p_2$$

where γ_F^q parameterizes v_F^q from $(+\infty, q_2)$ to $(-\infty, q_1)$ and γ_2 parameterizes v_2 from p_2 to q_2. Whereas, $(F_{21})_\square(\partial_k^{G_1}(gq_1))$

$$= (F_{21})_\square\left(\sum_{p_1 \in Cr_{k-1}(f_1)} \sum_{v_1 \in M_{f_1}(q_1,p_1)} \epsilon(v_1)(\gamma_1)_*(g)p_1\right)$$

$$= \sum_{\substack{p_1 \in Cr_{k-1}(f_1) \\ v_1 \in M_{f_1}(q_1,p_1)}} \epsilon(v_1)(F_{21})_\square\left((\gamma_1)_*(g)p_1\right)$$

$$= (-1)^{\lambda_{p_1}} \sum_{\substack{p_1 \in Cr_{k-1}(f_1) \\ v_1 \in M_{f_1}(q_1,p_1)}} \epsilon(v_1) \sum_{\substack{p_2 \in Cr_{k-1}(f_2) \\ v_F^p \in M_F(p_1,p_2)}} \epsilon(v_F^p)(\gamma_F^p)_*\left(\Phi((\gamma_1)_*(g))\right) p_2$$

$$= (-1)^{k-1} \sum_{\substack{p_1 \in Cr_{k-1}(f_1) \\ v_1 \in M_{f_1}(q_1,p_1)}} \sum_{\substack{p_2 \in Cr_{k-1}(f_2) \\ v_F^p \in M_F(p_1,p_2)}} \epsilon(v_1)\epsilon(v_F^p)((\pi \circ \gamma_F^p)\gamma_1)_*(\Phi(g))p_2$$

where γ_F^p parameterizes v_F^p from $(+\infty, p_2)$ to $(-\infty, p_1)$ and γ_1 parameterizes v_1 from p_1 to q_1.

Now recall that the boundary points of the path components of $\overline{M}_F(q_1, p_2)$ correspond to the terms in the above sums for $\partial_k^{G_2}((F_{21})_\square(gq_1))$ and $(F_{21})_\square(\partial_k^{G_1}(gq_1))$. Considering the three cases discussed before the statement of the proposition, we observe the following. The terms corresponding to boundary points in case (1) cancel each other out in the sum for $(F_{21})_\square(\partial_k^{G_1}(gq_1))$ because the homomorphism $((\pi \circ \gamma_F^p)\gamma_1)_*$ is constant on each path component and $\epsilon(\tilde{v}_1)\epsilon(\tilde{v}_F^p) + \epsilon(v_1)\epsilon(v_F^p) = 0$. Similarly, the terms corresponding to boundary points in case 3) cancel each other out in the sum for $\partial_k^{G_2}((F_{21})_\square(gq_1))$ because the homomorphism $(\gamma_2(\pi \circ \gamma_F^q))_*$ is constant on each path component and

$\epsilon(v_F^q)\epsilon(v_2) + \epsilon(\tilde{v}_F^q)\epsilon(\tilde{v}_2) = 0$. Finally, for the boundary points in case (2) we have

$$((\pi \circ \gamma_F^p)\gamma_1)_*(\Phi(g))p_2 = (\gamma_2(\pi \circ \gamma_F^q))_*(\Phi(g))p_2$$

$$\epsilon(v_1)\epsilon(v_F^p) = -\epsilon(v_F^q)\epsilon(v_2)$$

for each pair of endpoints in the same path component $[(v_1, v_F^p)] = [(v_F^q, v_2)]$. Therefore, $\partial_k((F_{21})_\square(q_1)) = (F_{21})_\square(\partial_k(q_1))$. □

Corollary 3.4 *The map $(F_{21})_\square$ induces a homomorphism in homology*

$$(F_{21})_* : H_k(C_*(f_1; G_1), \partial_*^{G_1}) \to H_k(C_*(f_2; G_2), \partial_*^{G_2})$$

for all $k = 0, \ldots, m$.

3.2 A Chain Homotopy

Assume that we have four Morse-Smale pairs (f_j, g_j) and four isomorphic bundles of abelian groups G_j, where $j = 1, 2, 3, 4$. For $j = 2, 3, 4$ there are families of isomorphisms $\Phi_{j1} : G_1 \to G_j$ making the diagram in Definition 2.2 commute, and if we define $\Phi_{42} = \Phi_{41} \circ \Phi_{21}^{-1}$, $\Phi_{43} = \Phi_{41} \circ \Phi_{31}^{-1}$, $\Phi_{ij} = \Phi_{ji}^{-1}$ when $i < j$, and $\Phi_{jj} = id$ for all j, then for all $i, j = 1, 2, 3, 4$ we have a family of isomorphisms $\Phi_{ji} : G_i \to G_j$ making the diagram in Definition 2.2 commute such that $\Phi_{42} \circ \Phi_{21} = \Phi_{41}$ and $\Phi_{43} \circ \Phi_{31} = \Phi_{41}$. Also, for all $i, j = 1, 2, 3, 4$ there are chain maps

$$(F_{ji})_\square : (C_*(f_i; G_i), \partial_*^{G_i}) \to (C_*(f_j; G_j), \partial_*^{G_j})$$

from Definition 3.1 and Proposition 3.3, defined using the family of isomorphisms Φ_{ji} and functions $F_{ji} : \overline{\mathbb{R}} \times M \to \mathbb{R}$.

We will show that under these assumptions the moduli spaces of gradient flow lines of a smooth function $H : \overline{\mathbb{R}} \times \overline{\mathbb{R}} \times M \to \mathbb{R}$ meeting certain transversality requirements can be used to construct a chain homotopy between the chain maps $(F_{42})_\square \circ (F_{21})_\square$ and $(F_{43})_\square \circ (F_{31})_\square$. By adding constants to f_1, f_2, and f_3 we may assume that $\inf f_1 > \sup f_2$, $\inf f_1 > \sup f_3$, $\inf f_2 > \sup f_4$, and $\inf f_3 > \sup f_4$.

Following [9] and [90] we let $H : \overline{\mathbb{R}} \times \overline{\mathbb{R}} \times M \to \mathbb{R}$ be a smooth function that is strictly decreasing in its first two components such that for some large $T \gg 0$ we have

$$H(s, t, x) = \begin{cases} f_1(x) - \rho(s) - \rho(t) & \text{if } s < -T \text{ and } t < -T \\ f_2(x) - \rho(s) - \rho(t) & \text{if } s > T \text{ and } t < -T \\ f_3(x) - \rho(s) - \rho(t) & \text{if } s < -T \text{ and } t > T \\ f_4(x) - \rho(s) - \rho(t) & \text{if } s > T \text{ and } t > T \end{cases}$$

3.2 A Chain Homotopy

where $\rho : \overline{\mathbb{R}} \to [-1, 1]$ is a smooth strictly increasing function with $\rho(-\infty) = -1$, $\rho(+\infty) = 1$, and $\lim_{t \to \pm\infty} \rho'(t) = 0$. The critical points of H are all on the boundary of $\overline{\mathbb{R}} \times \overline{\mathbb{R}} \times M$ since H is strictly decreasing in its first two components. Pick a Riemannian metric g on $\overline{\mathbb{R}} \times \overline{\mathbb{R}} \times M$ such that

$$\mathsf{g}(s, t, x) = \begin{cases} \mathsf{g}_1(x) + ds^2 + dt^2 & \text{if } s < -T \text{ and } t < -T \\ \mathsf{g}_2(x) + ds^2 + dt^2 & \text{if } s > T \text{ and } t < -T \\ \mathsf{g}_3(x) + ds^2 + dt^2 & \text{if } s < -T \text{ and } t > T \\ \mathsf{g}_4(x) + ds^2 + dt^2 & \text{if } s > T \text{ and } t > T \end{cases}$$

for all $x \in M$. The negative gradient flow of $H : \overline{\mathbb{R}} \times \overline{\mathbb{R}} \times M \to \mathbb{R}$ with respect to g can be pictured as follows, where M is in the vertical direction.

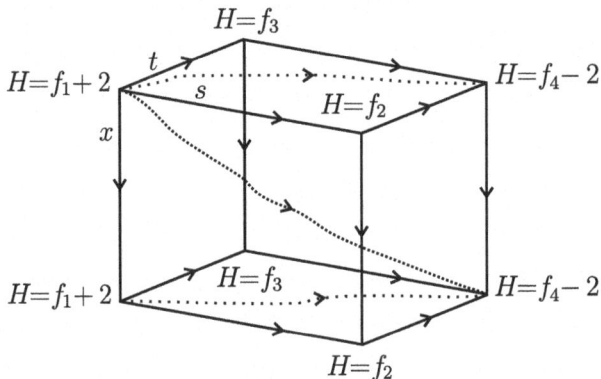

Let $\varphi_\tau : \overline{\mathbb{R}} \times \overline{\mathbb{R}} \times M \to \overline{\mathbb{R}} \times \overline{\mathbb{R}} \times M$ be the flow associated to $-\nabla H$, the negative gradient of H with respect to g, and for $p_1 \in Cr(f_1)$ and $q_j \in Cr(f_j)$ for some $j = 2, 3, 4$ define

$$W^u_H(p_1) = \{(s, t, x) \in \overline{\mathbb{R}} \times \overline{\mathbb{R}} \times M | \lim_{\tau \to -\infty} \varphi_\tau(s, t, x) = (-\infty, -\infty, p_1)\}$$

$$W^u_H(q_2) = \{(s, t, x) \in \overline{\mathbb{R}} \times \overline{\mathbb{R}} \times M | \lim_{\tau \to -\infty} \varphi_\tau(s, t, x) = (+\infty, -\infty, q_2)\}$$

$$W^s_H(q_2) = \{(s, t, x) \in \overline{\mathbb{R}} \times \overline{\mathbb{R}} \times M | \lim_{\tau \to +\infty} \varphi_\tau(s, t, x) = (+\infty, -\infty, q_2)\}$$

$$W^u_H(q_3) = \{(s, t, x) \in \overline{\mathbb{R}} \times \overline{\mathbb{R}} \times M | \lim_{\tau \to -\infty} \varphi_\tau(s, t, x) = (-\infty, +\infty, q_3)\}$$

$$W^s_H(q_3) = \{(s, t, x) \in \overline{\mathbb{R}} \times \overline{\mathbb{R}} \times M | \lim_{\tau \to +\infty} \varphi_\tau(s, t, x) = (-\infty, +\infty, q_3)\}$$

$$W^s_H(q_4) = \{(s, t, x) \in \overline{\mathbb{R}} \times \overline{\mathbb{R}} \times M | \lim_{\tau \to +\infty} \varphi_\tau(s, t, x) = (+\infty, +\infty, q_4)\}$$

$$W_H(p_1, q_j) = W_H^u(p_1) \cap W_H^s(q_j) \subset \overline{\mathbb{R}} \times \overline{\mathbb{R}} \times M$$
$$W_H(q_j, q_4) = W_H^u(q_j) \cap W_H^s(q_4) \subset \overline{\mathbb{R}} \times \overline{\mathbb{R}} \times M.$$

Since f_j satisfies the Morse-Smale transversality condition with respect to g for all $j = 1, 2, 3, 4$, we can choose H so that

- $W_H^u(p_1) \pitchfork W_H^s(q_j)$ for all $p_1 \in Cr(f_1)$ and $q_j \in Cr(f_j)$ for $j = 2, 3, 4$,
- $W_H^u(q_j) \pitchfork W_H^s(q_4)$ for all $q_j \in Cr(f_j)$ and $q_4 \in Cr(f_4)$ for $j = 2, 3$.

These conditions imply that H restricted to each face of $\overline{\mathbb{R}} \times \overline{\mathbb{R}} \times M$ defines a chain map. That is, if we define $F_{21}(s, x) = H(s, -\infty, x)$, $F_{42}(t, x) = H(+\infty, t, x)$, $F_{31}(t, x) = H(-\infty, t, x)$, and $F_{43}(s, x) = H(s, +\infty, x)$, then the chain maps $(F_{21})_\square$, $(F_{42})_\square$, $(F_{31})_\square$, $(F_{43})_\square$ from Definition 3.1 and Proposition 3.3 are defined. Conversely, given functions $F_{21}, F_{42}, F_{31}, F_{43}$ on the faces of $\overline{\mathbb{R}} \times \overline{\mathbb{R}} \times M$ that satisfy the transversality conditions needed to define the chain maps in Definition 3.1, we can choose an H meeting the above transversality conditions that agrees with these functions on the faces of $\overline{\mathbb{R}} \times \overline{\mathbb{R}} \times M$.

Choosing orientations for the unstable manifolds $W_{f_1}^u(p_1)$ and $W_{f_4}^u(q_4)$ then determines an orientation on $W_H(p_1, q_4)$ via the short exact sequence

$$T_* W_H(p_1, q_4) \hookrightarrow T_* W_H^u(p_1)|_{W_H(p_1,q_4)} \twoheadrightarrow \nu_*(W_H(p_1, q_4), W_H^u(p_1))|_{W_H(p_1,q_4)} \to 0$$

where the fibers of the normal bundle are isomorphic to $T_{q_4} W_{f_4}^u(q_4)$ via the flow of $-\nabla H$ and $W_H^u(q_1)$ is oriented as follows. For $\tau \ll 0$, $\varphi_\tau(s, t, x)$ will be in the region where $s, t < -T$. In that region the tangent space to $W_H^u(q_1)$ has the product orientation $\mathbb{R} \times \mathbb{R} \times T_* W_{f_1}^u(q_1)$, and this orientation then determines an orientation on $T_{(s,t,x)} W_H^u(q_1)$ via $\varphi_{-\tau}$. Taking a quotient by the action of $\mathbb{R}$ given by the negative gradient flow then gives an oriented smooth manifold

$$\mathcal{M}_H(p_1, q_4) = (W_H^u(p_1) \cap W_H^s(q_4))/\mathbb{R},$$

of dimension $\lambda_{p_1} - \lambda_{q_4} + 1$, where the orientation on $\mathcal{M}_H(p_1, q_4)$ is chosen by putting $-\nabla H$ first.

Definition 3.5 (Chain Homotopy) Let $f_j : M \to \mathbb{R}$ be smooth Morse-Smale functions on M and let G_j be isomorphic bundles of abelian groups over M, for $j = 1, 2, 3, 4$. Let G_4^* denote the pullback of G_4 under the projection $\pi : \overline{\mathbb{R}} \times \overline{\mathbb{R}} \times M \to M$, and let $\Phi_{41} : G_1 \to G_4$ be a family of isomorphisms making the diagram in Definition 2.2 commute. Assume that $H : \overline{\mathbb{R}} \times \overline{\mathbb{R}} \times M \to \mathbb{R}$ and the metric g on $\overline{\mathbb{R}} \times \overline{\mathbb{R}} \times M$ satisfy the conditions listed above, and define

3.2 A Chain Homotopy

$H_\square : (C_*(f_1; G_1), \partial_*^{G_1}) \to (C_*(f_4; G_4), \partial_*^{G_4})$ to be the linear map given on an elementary chain $gp_1 \in C_k(f_1; G_1)$ by

$$H_\square(gp_1) = \sum_{q_4 \in Cr_{k+1}(f_4)} \sum_{v_H \in \mathcal{M}_H(p_1,q_4)} \epsilon(v_H)(\gamma_H)_*(\Phi_{41}(g))q_4$$

where $\gamma_H : [0, 1] \to \overline{\mathbb{R}} \times \overline{\mathbb{R}} \times M$ is any continuous path from $(+\infty, +\infty, q_4)$ to $(-\infty, -\infty, p_1)$ such that the image $\gamma_H((0, 1))$ coincides with $Im(v_H)$, $\Phi_{41}(g) \in (G_4)_{p_1} = (G_4^*)_{(-\infty,-\infty,p_1)}$, $\epsilon(v_H) = \pm 1$ is the sign determined by the orientation on $\mathcal{M}_H(p_1, q_4)$, and λ_{p_1} denotes the index of p_1 as a critical point of $f_1 : M \to \mathbb{R}$.

There are compactification results for the moduli spaces of H similar to those stated in the previous section, cf. Section 2.4.4 of [78]. Hence, the moduli space $\mathcal{M}_H(p_1, q_4)$ has a compactification $\overline{\mathcal{M}}_H(p_1, q_4)$, consisting of all the piecewise gradient flow lines of H from $(-\infty, -\infty, p_1)$ to $(+\infty, +\infty, q_4)$. This implies that the above sum over $\mathcal{M}_H(p_1, q_4)$ is finite.

Proposition 3.6 *Let $f_j : M \to \mathbb{R}$ be smooth Morse-Smale functions on M and let G_j be isomorphic bundles of abelian groups over M, for $j = 1, 2, 3, 4$. The map $H_\square : (C_*(f_1; G_1), \partial_*^{G_1}) \to (C_*(f_4; G_4), \partial_*^{G_4})$ is a chain homotopy between $(F_{42})_\square \circ (F_{21})_\square$ and $(F_{43})_\square \circ (F_{31})_\square$, i.e.*

$$(F_{43})_\square \circ (F_{31})_\square - (F_{42})_\square \circ (F_{21})_\square = \partial_{k+1}^{G_4} H_\square + H_\square \partial_k^{G_1}$$

for all $k = 0, \ldots, m$.

Proof Choose families of isomorphisms $\Phi_{ji} : G_i \to G_j$ for $i, j = 1, 2, 3, 4$ making the diagram in Definition 2.2 commute such that $\Phi_{42} \circ \Phi_{21} = \Phi_{43} \circ \Phi_{31} = \Phi_{41}$. Let $q_1 \in Cr_k(f_1)$ and $q_4 \in Cr_k(f_4)$ for some $k = 0, \ldots, m$, where $m = \dim M$, and consider the compactified moduli space $\overline{\mathcal{M}}_H(q_1, q_4)$, where $H : \overline{\mathbb{R}} \times \overline{\mathbb{R}} \times M \to \mathbb{R}$ is a smooth function satisfying the transversality conditions listed before the proposition such that $H(s, -\infty, x) = F_{21}(s, x)$, $H(+\infty, t, x) = F_{42}(t, x)$, $H(-\infty, t, x) = F_{31}(t, x)$, and $H(s, +\infty, x) = F_{43}(s, x)$.

Since $\lambda_{q_1} = \lambda_{q_4}$, this compactified moduli space is an oriented compact smooth manifold with boundary of dimension one, and hence it is diffeomorphic to a disjoint union of oriented intervals. The endpoints of these intervals correspond to broken gradient flow lines of $H : \overline{\mathbb{R}} \times \overline{\mathbb{R}} \times M \to \mathbb{R}$ from $(-\infty, -\infty, q_1)$ to $(+\infty, +\infty, q_4)$, i.e. elements of the following zero dimensional spaces.

1. $\mathcal{M}_H(q_1, q_2) \times \mathcal{M}_H(q_2, q_4)$ some $q_2 \in Cr_k(f_2)$
2. $\mathcal{M}_H(q_1, q_3) \times \mathcal{M}_H(q_3, q_4)$ some $q_3 \in Cr_k(f_3)$
3. $\mathcal{M}_{f_1}(q_1, p_1) \times \mathcal{M}_H(p_1, q_4)$ some $p_1 \in Cr_{k-1}(f_1)$
4. $\mathcal{M}_H(q_1, r_4) \times \mathcal{M}_{f_4}(r_4, q_4)$ some $r_4 \in Cr_{k+1}(f_4)$

46   3 The Homology Determined by the Isomorphism Class of G

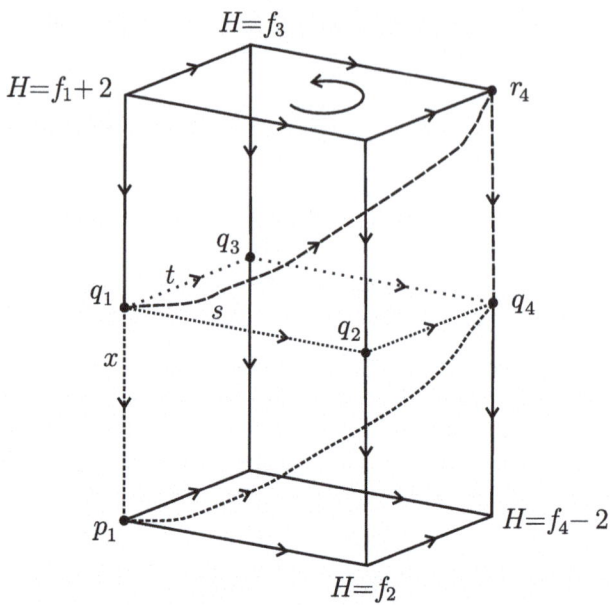

Now let $g_1 \in (G_1)_{q_1}$ and note that up to sign we have the following.

1. $((F_{42})_\square \circ (F_{21})_\square)(g_1 q_1)$ is defined by summing over elements of type 1.
2. $((F_{43})_\square \circ (F_{31})_\square)(g_1 q_1)$ is defined by summing over elements of type 2.
3. $(H_\square \partial_k^{G_1})(g_1 q_1)$ is defined by summing over elements of type 3.
4. $(\partial_{k+1}^{G_4} H_\square)(g_1 q_1)$ is defined by summing over elements of type 4.

However, these sums are all defined using signs determined by orientations, and the orientations on the moduli spaces of H aren't necessarily the same as the orientations on the moduli spaces of F_{ji} for $(i,j) = (1,2), (1,3), (2,4), (3,4)$.

Choose orientations for $W^u_{f_j}(q_j)$ for $j = 1, 2, 3, 4$. These orientations then determine orientations on $W^u_H(q_i)$ and $W^u_{F_{ji}}(q_i)$ for $(i,j) = (1,2), (1,3), (2,4), (3,4)$ using the product orientation and the gradient flow. Note that $W^u_H(q_2) = W^u_{F_{42}}(q_2)$, $W^u_H(q_3) = W^u_{F_{43}}(q_3)$, and $W^u_H(q_4) = W^u_{f_4}(q_4)$ as oriented manifolds, and we have the following dimensions.

$$\dim W^u_H(q_1) = k+2$$
$$\dim W^u_H(q_2) = \dim W^u_H(q_3) = k+1$$
$$\dim W^u_H(q_4) = k.$$

The spaces $W_H(q_i, q_j)$ are oriented using the short exact sequences

$$T_* W_H(q_i, q_j) \hookrightarrow T_* W^u_H(q_i)|_{W_H(q_i, q_j)} \twoheadrightarrow \nu_*(W_H(q_i, q_j), W^u_H(q_i))|_{W_H(q_i, q_j)} \to 0$$

3.2 A Chain Homotopy

where the fibers of the normal bundle are isomorphic to $T_{q_j} W_H^u(q_j)$, and the moduli space $\mathcal{M}_H(q_i, q_j) = W_H(q_i, q_j)/\mathbb{R}$ is oriented by putting $-\nabla H$ first. The spaces $W_{F_{ji}}(q_i, q_j)$ are oriented similarly using the short exact sequences

$$T_* W_{F_{ji}}(q_i, q_j) \hookrightarrow T_* W_{F_{ji}}^u(q_i)|_{W_{F_{ji}}(q_i,q_j)} \twoheadrightarrow \nu_*(W_{F_{ji}}(q_i, q_j), W_{F_{ji}}^u(q_i))|_{W_{F_{ji}}(q_i,q_j)} \to 0$$

where the fibers of the normal bundle are isomorphic to $T_{q_j} W_{f_j}^u(q_j)$, and the moduli space $\mathcal{M}_{F_{ji}}(q_i, q_j) = W_{F_{ji}}(q_i, q_j)/\mathbb{R}$ is oriented by putting $-\nabla F_{ji}$ first.

Since $W_H^u(q_2) = W_{F_{42}}^u(q_2)$ and $W_H^u(q_4) = W_{f_4}^u(q_4)$ as oriented manifolds, the above orientation conventions imply that

$$W_H(q_2, q_4) = W_{F_{42}}(q_2, q_4)$$

as oriented manifold. Similarly, $W_H^u(q_3) = W_{F_{43}}^u(q_3)$ and $W_H^u(q_4) = W_{f_4}^u(q_4)$ imply

$$W_H(q_3, q_4) = W_{F_{43}}(q_3, q_4).$$

Now, let $y = (s, -\infty, x) \in W_H^u(q_1)$ be in the region where $H(s, t, x) = f_1(x) - \rho(s) - \rho(t)$, and let $(v_1, v_2, w_1, w_2, \ldots, w_k)$ be a positive basis for $T_y W_H^u(q_1) = \mathbb{R} \times \mathbb{R} \times T_x W_{f_1}^u(q_1)$. Then $(v_2, w_1, w_2, \ldots, w_k)$ is a positive basis for $T_y W_{F_{21}}^u(q_1) = \mathbb{R} \times T_x W_{f_1}^u(q_1)$. If $(w_1', w_2', \ldots, w_k')$ is a positive basis for $T_{q_2} W_{f_2}^u(q_2)$, then $(v_2, w_1', w_2', \ldots, w_k')$ is a positive basis for $T_{q_2} W_H^u(q_2)$, where we have identified $v_2 \in T_y W_H^u(q_1)$ with its image under the map induced by the gradient flow in the s direction. The gradient flow in the s direction transports the basis $(w_1', w_2', \ldots, w_k')$ along a gradient flow line $\gamma \in W_{F_{21}}(q_1, q_2)$ to a basis for $\nu_y(W_{F_{21}}(q_1, q_2), W_{F_{21}}^u(q_1))$, which we still denote by $(w_1', w_2', \ldots, w_k')$. With this same identification, $(v_2, w_1', w_2', \ldots, w_k')$ is then a basis for $\nu_y(W_H(q_1, q_2), W_H^u(q_1))$. Thus, $\gamma \in W_{F_{21}}(q_1, q_2)$ is oriented in the v_1 direction if $(w_1, w_2, \ldots, w_k)$ and $(w_1', w_2', \ldots, w_k')$ determine the same orientation on $T_x W_{f_1}^u(q_1)$, since $(v_1, w_1', w_2', \ldots, w_k')$ will be a positive basis for $T_y W_{F_{21}}^u(q_1)$, and γ is oriented in the $-v_1$ direction if these orientations are opposite, because $(-v_1, w_1', w_2', \ldots, w_k')$ would then be a positive basis for $T_y W_{F_{21}}^u(q_1)$. Since the same is true for γ viewed as an element of $W_H(q_1, q_2)$, we have shown that

$$W_H(q_1, q_2) = W_{F_{21}}(q_1, q_2)$$

as oriented manifolds.

Finally, let $y = (-\infty, t, x) \in W_H^u(q_1)$ be in the region where $H(s, t, x) = f_1(x) - \rho(s) - \rho(t)$, and let $(v_1, v_2, w_1, w_2, \ldots, w_k)$ be a positive basis for $T_y W_H^u(q_1) = \mathbb{R} \times \mathbb{R} \times T_x W_{f_1}^u(q_1)$. Then $(v_2, w_1, w_2, \ldots, w_k)$ is a positive basis for $T_y W_{F_{31}}^u(q_1) = \mathbb{R} \times T_x W_{f_1}^u(q_1)$. If $(w_1', w_2', \ldots, w_k')$ is a positive basis for $T_{q_3} W_{f_3}^u(q_3)$, then $(v_1, w_1', w_2', \ldots, w_k')$ is a positive basis for $T_{q_3} W_H^u(q_3)$, where we have identified $v_1 \in T_y W_H^u(q_1)$ with its image under the map induced by the gradient flow in the t direction. The gradient flow in the t direction transports the basis $(w'_1, w'_2, \ldots, w'_k)$ along a gradient flow line $\gamma \in W_{F_{31}}(q_1, q_3)$ to a basis for $v_y(W_{F_{31}}(q_1, q_3), W^u_{F_{31}}(q_1))$, which we still denote by $(w'_1, w'_2, \ldots, w'_k)$. With this same identification, $(v_1, w'_1, w'_2, \ldots, w'_k)$ is then a basis for $v_y(W_H(q_1, q_3), W^u_H(q_1))$. Thus, $\gamma \in W_{F_{31}}(q_1, q_3)$ is oriented in the v_2 direction if $(w_1, w_2, \ldots, w_k)$ and $(w'_1, w'_2, \ldots, w'_k)$ determine the same orientation on $T_x W^u_{f_1}(q_1)$, and γ is oriented in the $-v_2$ direction if these orientations are opposite. On the other hand, γ viewed as an element of $W_H(q_1, q_3)$ is oriented in the $-v_2$ direction if $(w_1, w_2, \ldots, w_k)$ and $(w'_1, w'_2, \ldots, w'_k)$ determine the same orientation on $T_x W^u_{f_1}(q_1)$, since $(-v_2, v_1, w'_1, w'_2, \ldots, w'_k)$ will be a positive basis for $T_y W^u_H(q_1)$, and γ is oriented in the v_2 direction if these orientations are opposite, because $(v_2, v_1, w'_1, w'_2, \ldots, w'_k)$ would then be a positive basis for $T_y W^u_H(q_1)$. Hence,

$$W_H(q_1, q_3) = -W_{F_{31}}(q_1, q_3)$$

as oriented manifolds.

This shows that we have the following orientation relations between moduli spaces of broken flow lines.

$$\mathcal{M}_H(q_1, q_2) \times \mathcal{M}_H(q_2, q_4) = +\mathcal{M}_{F_{21}}(q_1, q_2) \times \mathcal{M}_{F_{43}}(q_2, q_4)$$
$$\mathcal{M}_H(q_1, q_3) \times \mathcal{M}_H(q_3, q_4) = -\mathcal{M}_{F_{31}}(q_1, q_3) \times \mathcal{M}_{F_{43}}(q_3, q_4)$$

To complete the proof, for $(i, j) = (1, 2), (1, 3), (2, 4), (3, 4)$ let $(H_{ji})_\square$ be the chain map defined using the same sum that defines $(F_{ji})_\square$ in Definition 3.1, but with the orientations from the moduli spaces $\mathcal{M}_{F_{ji}}(q_i, q_j)$ replaced with those of $\mathcal{M}_H(q_i, q_j)$. Since $\overline{\mathcal{M}}_H(q_1, q_4)$ is an oriented compact smooth one dimensional manifold with a coherent orientation on its boundary, summing over the moduli spaces of broken flow lines that make up the boundary of $\overline{\mathcal{M}}_H(q_1, q_4)$ gives

$$(H_{43})_\square \circ (H_{31})_\square + (H_{42})_\square \circ (H_{21})_\square + \partial^{G_4}_{k+1} H_\square + H_\square \partial^{G_1}_k = 0.$$

The above relations concerning the orientations of the moduli spaces of H versus those of F_{ji} for $(i, j) = (1, 2), (1, 3), (2, 4), (3, 4)$ imply that $(H_{42})_\square \circ (H_{21})_\square = (F_{42})_\square \circ (F_{21})_\square$ and $(H_{43})_\square \circ (H_{31})_\square = -(F_{43})_\square \circ (F_{31})_\square$. Therefore,

$$(F_{43})_\square \circ (F_{31})_\square - (F_{42})_\square \circ (F_{21})_\square = \partial^{G_4}_{k+1} H_\square + H_\square \partial^{G_1}_k.$$

This completes the proof of Proposition 3.6. □

3.3 An Invariance Theorem

We now prove the main theorem in this chapter.

Lemma 3.7 *Let $f_1 : M \to \mathbb{R}$ be a Morse-Smale function on $(M, \mathfrak{g}_1)$ and $\Phi : G_1 \to G_2$ a family of isomorphisms between two bundles of abelian groups G_1 and G_2 over M. For all $k = 0, \ldots, m$ the chain map*

$$(F_{21})_\square : C_k(f_1; G_1) \to C_k(f_2; G_2)$$

from Definition 3.1 is an isomorphism if we take $f_1 = f_2$, $\mathfrak{g}$ the product metric $\mathfrak{g}_1 + dt^2$ on $\overline{\mathbb{R}} \times M$, and $F_{21}(t, x) = f_1(x) - \rho(t)$. Moreover, on an elementary chain $gq \in C_k(f_1; G_1)$ we have $(F_{21})_\square(gq) = (-1)^k \Phi(g)q$. Thus, $(F_{12})_\square \circ (F_{21})_\square = id$ if $(F_{12})_\square$ is defined using the product metric, $F_{12}(t, x) = f_2(x) - \rho(t)$, and $\Phi^{-1} : G_2 \to G_1$.

Proof Note that $-\nabla F_{21} = (\rho'(t), -\nabla f_1)$ where $\rho'(t) > 0$ for all t since $\mathfrak{g}$ is the product metric. Recalling that $f_1 = f_2$ decreases along its gradient flow lines, we see that for any elementary chain $gq \in C_k(f_1; G_1)$ we have

$$\begin{aligned}(F_{21})_\square(gq) &= (-1)^{\lambda_q} \sum_{q_2 \in Cr_k(f)} \sum_{v_F \in M_F(q, q_2)} \epsilon(v_F)(\gamma_F)_*(\Phi(g))q_2 \\ &= (-1)^k \epsilon(v_F)(\gamma_F)_*(\Phi(g))q \\ &= (-1)^k \Phi(g)q,\end{aligned}$$

where $(\gamma_F)_* = id$ since $\pi \circ \gamma_F \equiv q$ and $\epsilon(v_F) = +1$ because for $\tau \ll 0$ the tangent bundle of $W_F^u(q)$ is oriented as $\mathbb{R} \times T_* W_{f_1}^u(q)$ and $W_F(q, q)$ is oriented by the relation

$$T_* W_F(q, q) \oplus T_q W_{f_2}^u(q) \approx T_* W_F^u(q)|_{W_F(q,q)}.$$

$\square$

Corollary 3.8 *For any two Morse-Smale pairs $(f_1, \mathfrak{g}_1)$ and $(f_2, \mathfrak{g}_2)$ and a family of isomorphisms $\Phi : G_1 \to G_2$ between bundles of abelian groups G_1 and G_2 over M, the time dependent gradient flow lines from f_1 to f_2 determine a canonical homomorphism*

$$(F_{21})_* : H_*((C_*(f_1; G_1), \partial_*^{G_1})) \to H_*((C_*(f_2; G_2), \partial_*^{G_2})),$$

i.e. the map $(F_{21})_*$ is independent of the choices made in the definition of $(F_{21})_\square$.

Proof Let $f_2 = f_3 = f_4$, $\mathfrak{g}_2 = \mathfrak{g}_3 = \mathfrak{g}_4$, and $G_2 = G_3 = G_4$ in Theorem 3.6, and let $F_{21} : \mathbb{R} \times M \to \mathbb{R}$ and $F_{31} : \mathbb{R} \times M \to \mathbb{R}$ be two functions that define time dependent gradient flow lines from f_1 to $f_2 = f_3 = f_4$ with respect to metrics $\mathfrak{g}_1$ and $\mathfrak{g}_2$ on $\overline{\mathbb{R}} \times M$ that are equal to $\mathfrak{g}_1 + dt^2$ and $\mathfrak{g}_2 + dt^2$ near the respective ends of $\overline{\mathbb{R}} \times M$. Theorem 3.6 implies

$$(F_{43})_* \circ (F_{31})_* = (F_{42})_* \circ (F_{21})_*$$

and Lemma 3.7 implies that $(F_{42})_*^{-1} = (F_{24})_*$ satisfies $(F_{42})_*^{-1} \circ (F_{43})_* = id$, if we define $(F_{24})_\square = (F_{34})_\square$ using the choices in Lemma 3.7. Thus, $(F_{31})_* = (F_{21})_*$. □

Theorem 3.9 (Invariance Theorem) *Let (M, g) be a closed finite dimensional smooth Riemannian manifold, and let G be a bundle of abelian groups over M. Then the homology of the twisted Morse-Smale-Witten chain complex $(C_*(f;G), \partial_*^G)$ is independent of the Morse-Smale pair (f, g) and depends only on the isomorphism class of the bundle of abelian groups G.*

Proof Let (f_1, g_1) and (f_2, g_2) be Morse-Smale pairs on M, and let $\Phi : G_1 \to G_2$ be a family of isomorphisms between bundles of abelian groups G_1 and G_2 over M. Letting $f_1 = f_3 = f_4$, $\mathsf{g}_1 = \mathsf{g}_3 = \mathsf{g}_4$, and $G_1 = G_3 = G_4$ in Theorem 3.6 we have

$$(F_{43})_* \circ (F_{31})_* = (F_{42})_* \circ (F_{21})_*$$

where $(F_{43})_* \circ (F_{31})_* = id$ by Lemma 3.7 and Corollary 3.8. Therefore, $(F_{12})_* \circ (F_{21})_* = id$. Similarly, $(F_{21})_* \circ (F_{12})_* = id$. □

Combining the preceding theorem with Proposition 2.6 we have the following.

Corollary 3.10 *Let $\eta \in \Omega_{cl}^1(M, \mathbb{R})$ be a closed one form on a Riemannian manifold (M, g). Then the homology of the η-twisted Morse-Smale-Witten chain complex $(C_*(f) \otimes \mathbb{R}, \partial_*^\eta)$ from Definition 2.12 is independent of the Morse-Smale pair (f, g) and depends only on the de Rham cohomology class of η.*

Chapter 4
Singular and CW-Homology with Local Coefficients

In this chapter we prove the Twisted Morse Homology Theorem, which says that a twisted Morse chain complex on a closed finite dimensional smooth manifold M with coefficients in a bundle of abelian groups G computes the singular homology of M with coefficients in G. This result is not surprising, but the proof is more technical than one might expect using CW homology. The main difficulty is that the CW-chain complex with local coefficients is not well-defined for general CW-complexes. To get a well-defined CW-chain complex with local coefficients one must restrict to a subclass of complexes, such as the regular CW-complexes. We prove a new theorem that shows that on a closed finite dimensional smooth manifold it is always possible to find a Riemannian metric and a Morse-Smale function whose unstable manifolds determine a regular CW-structure. In fact, it is always possible to find a Morse-Smale pair (f, g) whose unstable manifolds coincide with a smooth triangulation (Theorem 4.12). This new result is essential for the proof of the Twisted Morse Homology Theorem using classical techniques contained in this chapter, and it may also be of independent interest in combinatorial Morse theory.

Theorem 4.1 (Twisted Morse Homology Theorem) *Let $f : M \to \mathbb{R}$ be a smooth Morse-Smale function on a closed finite dimensional smooth Riemannian manifold (M, g), and let G be a bundle of abelian groups over M. The homology of the Morse-Smale-Witten chain complex with coefficients in G is isomorphic to the singular homology of M with coefficients in G, i.e.*

$$H_k((C_*(f; G), \partial_*^G)) \approx H_k(M; G)$$

for all $k = 0, \ldots, m$.

We will prove this theorem by observing that if the unstable manifolds of the Morse-Smale function $f : M \to \mathbb{R}$ determine a **regular** CW-structure on (M, g), then the Morse-Smale-Witten chain complex $(C_*(f; G), \partial_*^G)$ and Steenrod's CW-chain complex with coefficients in the bundle of abelian groups G are isomorphic

(Lemma 4.10). We will then show how to construct a Morse-Smale pair (f, g) whose unstable manifolds determine a regular CW-structure on M (Theorem 4.12) and apply the above Invariance Theorem (Theorem 3.9). This proves the Twisted Morse Homology Theorem because the homology of Steenrod's CW-chain complex with coefficients in G (for a regular CW-complex) is isomorphic to the singular homology of M with coefficients in G (Lemma 4.8).

4.1 Singular Homology with Local Coefficients

For the convenience of the reader we now recall the definition of singular homology with coefficients in a bundle of abelian groups. For more details see Chapter VI of [93].

Let G be a bundle of abelian groups over a topological space X. Let Δ^k denote the standard k-simplex with vertices $e_0, \ldots, e_k$, and let $C_k(X; G)$ be the set of all functions c such that the following conditions hold.

1. For every singular k-simplex $u : \Delta^k \to X$, $c(u) \in G_{u(e_0)}$ is defined.
2. The set of singular simplices u such that $c(u) \neq 0$ is finite.

Elements of the abelian group $C_k(X; G)$ are called **singular k-chains with coefficients in G**, and every $c \in C_k(X; G)$ can be represented as a finite sum

$$c = \sum_{i=1}^{n} c(u_i) \cdot u_i$$

where $u_1, \ldots, u_n$ are the singular simplices such that $c(u_i) \neq 0$ and $c(u_i) \in G_{u_i(e_0)}$ for all $i = 1, \ldots, n$. The **relative singular k-chains with coefficients in G** for a subspace $A \subseteq X$, denoted $C_k(X, A; G)$, are defined similarly.

Definition 4.2 The **singular boundary operator with coefficients in G** is defined to be the homomorphism $\partial_k : C_k(X; G) \to C_{k-1}(X; G)$ given on an elementary chain $c = g \cdot u$ by

$$\partial_k(g \cdot u) = (\gamma_u)_*(g) \cdot u \circ F_0 + \sum_{i=1}^{k}(-1)^i g \cdot u \circ F_i$$

where $(\gamma_u)_* : G_{u(e_0)} \to G_{u(e_1)}$ is the homomorphism associated to the path $\gamma_u(t) = u((1-t)e_1 + te_0)$ from $u(e_1)$ to $u(e_0)$ and $F_i : \Delta^{k-1} \hookrightarrow \Delta^k$ is the inclusion onto the face opposite e_i for all $i = 0, \ldots, k-1$. The pair $(C_*(X; G), \partial_*)$ is a chain complex, and its homology groups $H_*(X; G)$ are called the **homology groups of X with coefficients in the bundle G**. The relative homology groups $H_*(X, A; G)$ for a subspace $A \subseteq X$ are defined similarly.

Eilenberg showed that singular homology with local coefficients is closely related to equivariant homology. Suppose that (X, x_0) is a connected topological space and G_0 is an abelian group on which $\pi_1(X, x_0)$ acts. Since $\pi_1(X, x_0)$ also acts on the universal cover $\widetilde{X}$ and this action commutes with the boundary operator on singular chains in $\widetilde{X}$, there is a chain complex $(G_0 \otimes_{\pi_1} C_*(\widetilde{X}), \bar{\partial}_*)$, where the tensor product is taken over $\pi_1(X, x_0)$ and the boundary operator $\bar{\partial}_*$ is induced from the boundary operator on the singular chains in $\widetilde{X}$. The homology groups of this complex are the equivariant homology groups $E_*(\widetilde{X}; G_0)$, cf. Sect. 2.6.

Let G be a bundle of abelian groups on X in the isomorphism class determined by the action of $\pi_1(X, x_0)$ on G_0 (Theorem 2.3). We have the following.

Theorem 4.3 (Eilenberg) *For all k, $H_k(X; G)$ is isomorphic to $E_k(\widetilde{X}; G_0)$.*

For a proof of the preceding theorem see Section VI.3 of [93].

4.2 Regular CW-Complexes

We now recall the definition of a regular CW-complex, and we prove that for a regular CW-complex Steenrod's cellular boundary operator with coefficients in a bundle of abelian groups G (Definition 4.6) is induced from the singular boundary operator with coefficients in G (Lemma 4.8).

It is important to note that Steenrod's cellular boundary operator with local coefficients is not well-defined for a general CW-complex. Originally, Steenrod defined his cellular boundary operator for topological spaces with a cellular decomposition, where the space is decomposed into cells homeomorphic to closed simplices (Section 19 of [84]). A cellular decomposition typically requires more cells than a CW-decomposition but fewer cells than a simplicial decomposition. With regard to the cellular boundary operator with local coefficients Steenrod stated, "For the generalization to be given, it seems to be necessary that the closed cells be at least simply connected" (Section 31 of [84]).

Thus, Steenrod's CW-chain complex with local coefficients is only defined for some restricted class of CW-complexes; although it is possible to define a CW-chain complex with local coefficients for a general CW-complex using a boundary operator induced from a connecting homomorphism (Section VI.4 of [93]).

Definition 4.4 A CW-complex X is **regular** if every closed k-cell e^k, with $k > 0$, is homeomorphic to Δ^k.

Remark 4.5 Not every CW-complex is regular. For instance, the CW-structure on S^1 given by the Morse function in Example 2.14 is not regular. In that example the closed 1-cell e^1 is homeomorphic to S^1 instead of Δ^1. Similarly, the CW-structure on the 2-torus with 4 cells is not regular, cf. Example 7.11 of [8].

Regular cell complexes satisfy the following properties, which are not necessarily satisfied by nonregular CW-complexes. For more details see Section IX.6 of [51] or Section II.6 of [93].

1. If e^k is a k-cell, then its **boundary** $\dot{e}^k = e^k - \text{int}(e^k)$ is the union of finitely many $(k-1)$-cells. (Note that the "boundary" in this context may be different than the topological boundary of e^k.)
2. If $j < k$ and e^j and e^k are cells such that $e^j \cap \dot{e}^k \neq \emptyset$, then $e^j \subset \dot{e}^k$.
3. For any k-cell e^k of X with $k \geq 0$, e^k and $\dot{e}^k$ are the underlying spaces of subcomplexes of X.
4. If e^k and e^{k+2} are cells such that e^k is a face of e^{k+2}, then there are exactly two $(k+1)$-cells e^{k+1} such that e^k is a proper face of e^{k+1} and e^{k+1} is a proper face of e^{k+2}, i.e. $e^k < e^{k+1} < e^{k+2}$.
5. The incidence number $[e^k : e^{k-1}]$ is ± 1 if $e^{k-1} < e^k$ and zero otherwise.

For any system of local coefficients G over a CW-complex X (not necessarily regular) the triple $(X^{(k-2)}, X^{(k-1)}, X^{(k)})$, where $X^{(k)}$ denotes the k-skeleton of X, determines a connecting homomorphism

$$H_k(X^{(k)}, X^{(k-1)}; G) \xrightarrow{\delta_k} H_{k-1}(X^{(k-1)}; G)$$

that can be composed with the map $H_{k-1}(X^{(k-1)}; G) \xrightarrow{j_*} H_{k-1}(X^{(k-1)}, X^{(k-2)}; G)$ induced from the inclusion $j : X^{(k-2)} \hookrightarrow X^{(k-1)}$ to give a map

$$H_k(X^{(k)}, X^{(k-1)}; G) \xrightarrow{\tilde{\partial}_k} H_{k-1}(X^{(k-1)}, X^{(k-2)}; G).$$

The above map satisfies $\tilde{\partial}_{k-1} \circ \tilde{\partial}_k = 0$, and the homology groups of the chain complex with boundary operator $\tilde{\partial}_k$ and kth-chain group $H_k(X^{(k)}, X^{(k-1)}; G)$ are isomorphic to the singular homology groups of X with coefficients in the bundle G from Definition 4.2, cf. Theorems VI.2.4 and VI.4.4 of [93].

Now if X is regular, then the homology group $H_k(X^{(k)}, X^{(k-1)}; G)$ can be represented as a direct sum of the images of the maps induced by the characteristic maps $f_\sigma : (\Delta^k, \dot{\Delta}^k) \to (e_\sigma^k, \dot{e}_\sigma^k) \subseteq (X^{(k)}, X^{(k-1)})$ of the k-cells e_σ^k of X as follows. For each k-cell e_σ^k we choose a basepoint $x(e_\sigma^k)$, which determines an isomorphism

$$\bigoplus_\sigma (f_\sigma)_* : \bigoplus_\sigma H_k(\Delta^k, \dot{\Delta}^k; G_{x(e_\sigma^k)}) \xrightarrow{\approx} H_k(X^{(k)}, X^{(k-1)}; G),$$

cf. Theorem VI.4.1 of [93]. Note that the definition of the induced map $(f_\sigma)_*$ requires both a map of spaces $f_\sigma : (\Delta^k, \dot{\Delta}^k) \to (X^{(k)}, X^{(k-1)})$ and a homomorphism $\gamma_* : G_{x(e_\sigma^k)} \to f_\sigma^*(G)$, cf. Section VI.2 of [93]. We take the homomorphism γ_* to be the one defined by restricting the local coefficient system G to the simply connected space e_σ^k, i.e. for any point $x \in \Delta^k$ there is a unique homotopy class of paths rel endpoints from $f_\sigma(x)$ to $x(e_\sigma^k)$ and hence a well-defined homomorphism

4.2 Regular CW-Complexes

$G_{x(e_\sigma^k)} \to G_{f_\sigma(x)}$. Since $G_{x(e_\sigma^k)} \approx H_k(\Delta^k, \dot{\Delta}^k; G_{x(e_\sigma^k)})$, we can use the above isomorphisms to identify

$$CW_k(X; G) \stackrel{\text{def}}{=} \left\{ \sum_\sigma g e_\sigma^k \,\middle|\, g \in G_{x(e_\sigma^k)} \right\}$$

$$= \bigoplus_\sigma G_{x(e_\sigma^k)}$$

$$= \bigoplus_\sigma H_k(\Delta^k, \dot{\Delta}^k; G_{x(e_\sigma^k)})$$

$$\stackrel{\oplus_\sigma (f_\sigma)_*}{=} H_k(X^{(k)}, X^{(k-1)}; G)$$

Definition 4.6 **Steenrod's cellular boundary operator with coefficients in G** for a regular CW-complex X with a system of local coefficients G is defined to be the homomorphism $\partial_k : CW_k(X; G) \to CW_{k-1}(X; G)$ given on an elementary chain ge^k by

$$\partial_k(ge^k) = \sum_{e^{k-1} < e^k} [e^k : e^{k-1}](\gamma_{e^{k-1}e^k})_*(g) e^{k-1},$$

where $(\gamma_{e^{k-1}e^k})_* : G_{x(e^k)} \to G_{x(e^{k-1})}$ denotes the isomorphism determined by any path from $x(e^{k-1})$ to $x(e^k)$ contained in the closure of e^k. We will call the pair $(CW_*(X; G), \partial_*)$ **Steenrod's CW-chain complex with coefficients in the bundle G**.

Note that $(\gamma_{e^{k-1}e^k})_*$ does not depend on the path from $x(e^{k-1})$ to $x(e^k)$ since $e^k \approx \Delta^k$ is simply connected. Moreover, one can show directly that $\partial_{k-1} \circ \partial_k = 0$, and the homology of Steenrod's CW-chain complex for a regular CW-complex X with coefficients in G is independent of the choice of basepoints $x(e^k)$; see [23, 52] or Section 31.2 of [84] for more details.

Remark 4.7 If the CW-complex X is not regular, then the homomorphism $(\gamma_{e^{k-1}e^k})_*$ may not be well-defined. For instance, the above formula for $\partial_1(ge^1)$ is not well-defined for the CW-structure given by the unstable manifolds of the height function on S^1 in Example 2.14. Similarly, the above formula is not well-defined for the CW-structure on the 2-torus with 4 cells, cf. Example 7.11 of [8].

Lemma 4.8 *If X is a regular CW-complex and G is a bundle of abelian groups over X, then the singular boundary operator with coefficients in G from Definition 4.2 induces Steenrod's cellular boundary operator with coefficients in G from Definition 4.6. That is, the following diagram commutes.*

$$CW_k(X; G) \xrightarrow{\partial_k} CW_{k-1}(X; G)$$

$$\updownarrow \qquad\qquad \updownarrow$$

$$H_k(X^{(k)}, X^{(k-1)}; G) \xrightarrow{\tilde{\partial}_k} H_{k-1}(X^{(k-1)}, X^{(k-2)}; G)$$

Thus, the homology of Steenrod's CW-chain complex $(CW_*(X; G), \partial_*)$ is isomorphic to the singular homology of X with coefficients in the bundle G.

Proof The connecting homomorphism δ_k is natural, cf. Theorem VI.2.4 of [93]. Hence, the following diagram commutes

$$\begin{array}{ccc}
H_k(\Delta^k, \dot{\Delta}^k; G_{x(e_\sigma^k)}) & \xrightarrow{\bar{\delta}_k} & H_{k-1}(\dot{\Delta}^k; G_{x(e_\sigma^k)}) \\
\downarrow {(f_\sigma)_*} & & \downarrow {(f_{\partial\sigma})_*} \\
H_k(X^{(k)}, X^{(k-1)}; G) & \xrightarrow{\delta_k} & H_{k-1}(X^{(k-1)}; G) & \xrightarrow{j_*} & H_{k-1}(X^{(k-1)}, X^{(k-2)}; G) \\
& & & & \approx \uparrow \oplus_\tau (f_\tau)_* \\
& & \oplus_\tau H_{k-1}\left(\Delta^{k-1}, \dot{\Delta}^{k-1}; G_{x(e_\tau^{k-1})}\right)
\end{array}$$

where f_σ and f_τ are characteristic maps, $f_{\partial\sigma} = f_\sigma|_{\dot{\Delta}^k}$, the homomorphisms $G_{x(e_\sigma^k)} \to f_\sigma^* G$ and $G_{x(e_\sigma^k)} \to f_{\partial\sigma}^* G$ are defined by restricting the local coefficient system to the simply connected space e_σ^k, the homomorphism $G_{x(e_\tau^{k-1})} \to f_\tau^* G$ is defined by restricting the local system to the simply connected space e_τ^{k-1}, and $j : (X^{(k-1)}, \emptyset) \to (X^{(k-1)}, X^{(k-2)})$ is the inclusion. Moreover, we have fixed isomorphisms

$$G_{x(e_\sigma^k)} = H_k(\Delta^k, \dot{\Delta}^k; G_{x(e_\sigma^k)})$$

and

$$\bigoplus_\tau H_{k-1}\left(\Delta^{k-1}, \dot{\Delta}^{k-1}; G_{x(e_\tau^{k-1})}\right) = CW_{k-1}(X; G).$$

The statement of the lemma claims that these isomorphisms and the map given by tracing the above diagram from the upper left to the lower right agree with ∂_k from Definition 4.6.

If we let $X^{(k-1)}/X^{(k-2)}$ be the space obtained by identifying $X^{(k-2)}$ to a point when $k > 1$ or $X^{(0)}$ union a disjoint basepoint $*$ when $k = 1$, then $X^{(k-1)}/X^{(k-2)}$ is a bouquet of $(k-1)$-spheres S_τ^{k-1} unioned at the basepoint $*$. For each τ let p_τ be the composition $X^{(k-1)} \to X^{(k-1)}/X^{(k-2)} \to S_\tau^{k-1}$, where the second map collapses

4.2 Regular CW-Complexes

every sphere other than S_τ^{k-1} to the basepoint. Then $p_\tau : (X^{(k-1)}, X^{(k-2)}) \to (S_\tau^{k-1}, *)$ is a map of pairs, and $p_\tau \circ f_{\tau'} : \Delta^{k-1} \to S_\tau^{k-1}$ is a constant map to the basepoint if $\tau \neq \tau'$ and the map that identifies the boundary of Δ^{k-1} to the basepoint when $\tau = \tau'$. Identifying $H_{k-1}(S_\tau^{k-1}, *; G_{x(e_\tau^{k-1})}) = H_{k-1}(\Delta^{k-1}, \dot{\Delta}^{k-1}; G_{x(e_\tau^{k-1})})$ we see that

$$\bigoplus_\tau (p_\tau)_* : H_{k-1}(X^{(k-1)}, X^{(k-2)}; G) \xrightarrow{\approx} \bigoplus_\tau H_{k-1}(\Delta^{k-1}, \dot{\Delta}^{k-1}; G_{x(e_\tau^{k-1})})$$

is the inverse of $\bigoplus_\tau (f_\tau)_*$, cf. Proposition 2.14 of [8].

Now, the characteristic maps $f_\sigma : \Delta^k \to e_\sigma^k$ and $f_\tau : \Delta^{k-1} \to e_\tau^{k-1}$ determine orientations on the cells e_σ^k and e_τ^{k-1}, and the sign $[e_\sigma^k : e_\tau^{k-1}] = \pm 1$ keeps track of the compatibility of these orientations. That is, if $e_\tau^{k-1} < e_\sigma^k$, then $[e_\sigma^k : e_\tau^{k-1}]$ is $+1$ or -1 depending on whether the following isomorphism preserves or reverses the orientations determined by f_σ and f_τ, where the last isomorphism is given by excision, cf. Section II.6 of [93].

$$H_k(e_\sigma^k, \dot{e}_\sigma^k) \xrightarrow{\bar{\delta}_k} H_{k-1}(\dot{e}_\sigma^k) \xrightarrow{j_*} H_{k-1}(\dot{e}_\sigma^k, \overline{\dot{e}_\sigma^k - e_\tau^{k-1}}) \xrightarrow{\approx} H_{k-1}(e_\tau^{k-1}, \dot{e}_\tau^{k-1})$$

Moreover, the exact sequence

$$H_k(\Delta^k; G_{x(e_\sigma^k)}) \to H_k(\Delta^k, \dot{\Delta}^k; G_{x(e_\sigma^k)}) \xrightarrow{\bar{\delta}_k} H_{k-1}(\dot{\Delta}^k; G_{x(e_\sigma^k)}) \to H_{k-1}(\Delta^k; G_{x(e_\sigma^k)})$$

shows that $\bar{\delta}_k$ is an isomorphism when $k > 1$ and injective when $k = 1$, where the orientation of $\dot{\Delta}^k$ is determined by the orientation of Δ^k. Thus, by the Universal Coefficient Theorem we have

$$H_k(\Delta^k, \dot{\Delta}^k; G_{x(e_\sigma^k)}) \approx \mathbb{Z} \otimes_\mathbb{Z} G_{x(e_\sigma^k)} \approx G_{x(e_\sigma^k)} \quad \text{for all } k \geq 0,$$

$$H_{k-1}(\dot{\Delta}^k; G_{x(e_\sigma^k)}) \approx \begin{cases} G_{x(e_\sigma^k)} & \text{if } k > 1 \\ G_{x(e_\sigma^k)} \oplus G_{x(e_\sigma^k)} & \text{if } k = 1, \end{cases}$$

and

$$\bar{\delta}_k(g) = \begin{cases} g & \text{if } k \geq 1 \\ (g, -g) & \text{if } k = 1, \end{cases}$$

cf. Example 2.2 of [8]. Since $p_\tau \circ j \circ f_{\partial\sigma} : \dot{\Delta}^{k-1} \to S_\tau^{k-1}$ is a map of degree $[e_\sigma^k : e_\tau^{k-1}]$, this shows that for any $g \in G_{x(e_\sigma^k)}$ we have

$$(p_\tau)_*(j_*(\delta_k((f_\sigma)_*(g)))) = (p_\tau)_*(j_*((f_{\partial\sigma})_*\bar{\delta}_k((g))))$$
$$= [e_\sigma^k : e_\tau^{k-1}](\gamma_{e_\sigma^k e_\tau^{k-1}})_*(g) \in G_{x(e_\tau^{k-1})}.$$

□

4.3 Unstable Manifolds and Regular CW-Structures

The unstable manifolds of the Morse-Smale functions in Examples 2.14 and 2.15 do not determine regular CW-structures on S^1 or $\mathbb{R}P^2$. This can be seen directly from Definition 4.4 or by noting that, in both cases, there are cells e^k and e^{k-1} with $e^{k-1} < e^k$ where the incidence number $[e^k : e^{k-1}] \neq \pm 1$. However, we will show in Theorem 4.12 that on any closed finite dimensional smooth manifold M it is always possible to find a Morse function $f : M \to \mathbb{R}$ and a Riemannian metric g such that f satisfies the Morse-Smale transversality condition with respect to g and the unstable manifolds of (f, g) determine a regular CW-structure on M.

Example 4.9 (A Deformed Circle) Consider the unit circle $M = S^1$ and the Morse-Smale function $f : S^1 \to \mathbb{R}$ given by the height function in the following picture, where the arrows indicate the orientations of the unstable manifolds. It's clear that the unstable manifolds of f determine a regular CW-structure on S^1 since $\overline{W^u(q_1)} = W^u(q_1) \cup \{p_1, p_2\} \approx \Delta^1$ and $\overline{W^u(q_2)} = W^u(q_2) \cup \{p_1, p_2\} \approx \Delta^1$.

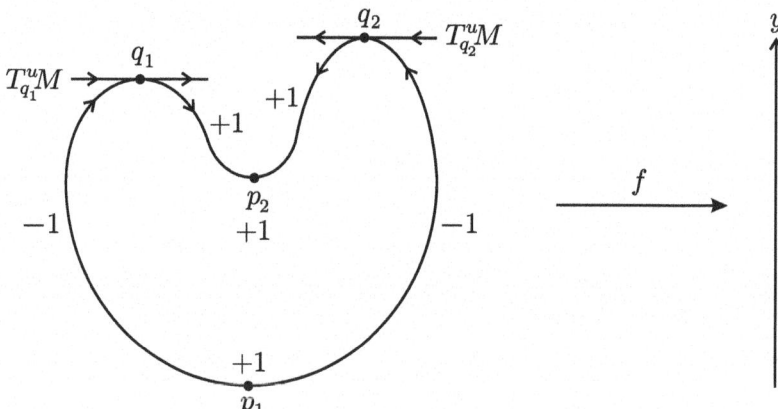

The height function on a deformed circle

The (untwisted) Morse-Smale-Witten complex of f is

$$
\begin{array}{ccccccc}
0 & \longrightarrow & C_1(f) & \xrightarrow{\partial_1} & C_0(f) & \longrightarrow & 0 \\
& & \updownarrow \approx & & \updownarrow \approx & & \\
0 & \longrightarrow & <q_1, q_2> & \xrightarrow{\partial_1} & <p_1, p_2> & \longrightarrow & 0
\end{array}
$$

with $\partial_1(q_1) = \partial_1(q_2) = p_2 - p_1$. The homology group $H_1(C_*(f), \partial_*) \approx \mathbb{Z}$ is generated by the class $[q_1 - q_2]$, and the image of ∂_1 is the group $<p_2 - p_1>$. Hence, $H_0(C_*(f), \partial_*) \approx <p_1, p_2> / <p_2 - p_1> \approx \mathbb{Z}$.

4.3 Unstable Manifolds and Regular CW-Structures

Now, let G be a bundle of abelian groups over S^1, and let $\gamma_{p_i q_j} : [0, 1] \to M$ be a path from p_i to q_j whose image coincides with the image of a gradient flow line $v_{q_j p_i}$, for $i, j = 1, 2$. The twisted Morse-Smale-Witten boundary operator is determined by

$$\partial_1^G(g_1 q_1) = (\gamma_{p_2 q_1})_*(g_1) p_2 - (\gamma_{p_1 q_1})_*(g_1) p_1 \quad \text{for all } g_1 \in G_{q_1}$$
$$\partial_1^G(g_2 q_2) = (\gamma_{p_2 q_2})_*(g_2) p_2 - (\gamma_{p_1 q_2})_*(g_2) p_1 \quad \text{for all } g_2 \in G_{q_2}.$$

Note that in this example it is not immediately obvious that the homology of the twisted chain complex $(C_*(f; G); \partial_*^G)$ depends only on the isomorphism class of G (Theorem 3.9). On the other hand, since the unstable manifolds of the Morse-Smale function determine a regular CW-structure $X^{(0)} \subseteq X^{(1)}$ on S^1, we can compare the twisted Morse-Small-Witten chain complex $(C_*(f; G); \partial_*^G)$ to Steenrod's twisted CW-chain complex $(CW_*(X; G), \partial_*)$.

If we choose the critical point $q_j = x(e_j^1)$ as the basepoint for the cell $e_j^1 = \overline{W^u(q_j)}$ for $j = 1, 2$, then the homomorphism $(\gamma_{e_i^0 e_j^1})_* : G_{q_j} \to G_{p_i}$ from Definition 4.6 coincides with $(\gamma_{p_i q_j})_*$ for $i, j = 1, 2$. Moreover, one can check directly that the signs $\epsilon(v_{q_j p_i})$ associated to the gradient flow lines agree with the corresponding incidence numbers $[e_j^1 : e_i^0]$. Hence, Steenrod's cellular boundary operator with coefficients in G (Definition 4.6) is the same as the Morse-Smale-Witten boundary operator with coefficients in G (Definition 2.11) once we identify the corresponding generators. Therefore, $H_k((C_*(f; G), \partial_*^G)) \approx H_*((CW_*(X; G), \partial_*)) \approx H_k(S^1; G)$ for all k.

Lemma 4.10 *Let $f : M \to \mathbb{R}$ be a Morse-Smale function on a closed finite dimensional smooth Riemannian manifold (M, g). Assume that the unstable manifolds of (f, g) determine a regular CW-structure X on M, and for each closed k-cell e_q^k with $W^u(q) \subset e_q^k$ choose the basepoint $x(e_q^k)$ to be the critical point $q \in Cr_k(f)$. Let G be a bundle of abelian groups over M, and identify the generator e_q^k in the CW-chain complex $(CW_*(X; G), \partial_*)$ with the generator q in the Morse-Smale-Witten chain complex $(C_*(f; G), \partial_*^G)$.*

Under these identifications, Steenrod's cellular boundary operator with coefficients in G from Definition 4.6 coincides with the twisted Morse-Smale-Witten boundary operator from Definition 2.11. Thus, $H_k((CW_*(X; G), \partial_*)) \approx H_k((C_*(f; G), \partial_*^G))$ for all $k = 0, \ldots, m$.

Proof The assumption that the CW-complex is regular implies that any two gradient flow lines from $q \in Cr_k(f)$ to $p \in Cr_{k-1}(f)$ are homotopic rel endpoints. Hence, if γ^v is any continuous path from p to q whose images coincide with $v \in \mathcal{M}(q, p)$, then the isomorphism $(\gamma^v)_* : G_q \to G_p$ is independent of v. That is, in the notation of Definition 4.6, we have $(\gamma^v)_* = (\gamma_{e_p^{k-1} e_q^k})_*$. Thus, for the twisted Morse-Smale-Witten boundary operator from Definition 2.11 we have for any $g \in G_q$

$$\partial_k^G(gq) = \sum_{p \in Cr_{k-1}(f)} \sum_{v \in \mathcal{M}(q,p)} \epsilon(v) \gamma_*^v(g) p$$

$$= \sum_{p \in Cr_{k-1}(f)} \left(\sum_{v \in \mathcal{M}(q,p)} \epsilon(v) \right) (\gamma_{e_p^{k-1} e_q^k})_*(g) p$$

$$= \sum_{e_p^{k-1} < e_q^k} \#\mathcal{M}(q,p) (\gamma_{e_p^{k-1} e_q^k})_*(g) p,$$

where the last equality follows from the fact that

$$\overline{W^u(q)} = \bigcup_{q \succeq p} W^u(p),$$

cf. Corollary 6.27 of [8], and

$$\#\mathcal{M}(q,p) \stackrel{\text{def}}{=} \sum_{v \in \mathcal{M}(q,p)} \epsilon(v).$$

Now, if the Riemannian metric g is locally trivial with respect to the Morse charts of f, i.e. if one can choose the Morse charts to be isometries with respect to the standard Euclidean metric on $\mathbb{R}^m$ (see Definition 2.16 of [66]), then by Theorem 3.9 of [66] for any $q \in Cr_k(f)$ and $p \in Cr_{k-1}(f)$ we have $\#\mathcal{M}(q,p) = [e_q^k : e_p^{k-1}]$ where $[e_q^k : e_p^{k-1}]$ is the incidence number used to define the CW-boundary operator. The same result was shown to hold for a general Morse-Smale metric in Theorem 9.3 of [67]. This proves the lemma since Steenrod's cellular boundary operator from Definition 4.6 is given by

$$\partial_k(g e_q^k) = \sum_{e_p^{k-1} < e_q^k} [e_q^k : e_p^{k-1}] (\gamma_{e_p^{k-1} e_q^k})_*(g) e_p^{k-1}.$$

□

Remark 4.11 (A Fundamental Identity) The proof of the previous lemma relied on the important identity

$$\#\mathcal{M}(q,p) = [e_q^k : e_p^{k-1}].$$

As noted in the proof, this identity was explicitly proved by Qin in [66] under the assumption that the metric g is locally trivial with respect to the Morse charts of f and then later extended to all metrics in [67]. It was also proved by Audin and Damian for gradient-like vector fields in Appendix 4.9 of [4]. The proof of the identity relies on the manifolds with corners structure on the compactified moduli space $\overline{\mathcal{M}}(q, p)$, which has been established by several authors under various assumptions on the Riemannian metric or the form of the gradient vector field in the Morse charts [4, 18, 19, 48, 66, 67, 91].

In Theorem 4.12 we show that it is always possible to find a Morse-Smale pair (f, g) on M for which there are Morse charts that are isometries with respect to the standard Euclidean metric on $\mathbb{R}^m$ around every critical point and the unstable manifolds of (f, g) determine a regular CW-structure on M. Moreover, the above identity is easy to establish directly for the function constructed in Theorem 4.12 (Remark 4.16). Thus, the proof of Theorem 4.1 is independent of the more general results proved in the above referenced papers.

4.4 A Morse-Smale Function that Determines a Regular CW-Structure

In the appendix to [13] Laudenbach proved that if M is a finite dimensional closed smooth manifold and $f : M \to \mathbb{R}$ is Morse-Smale with respect to a Riemannian metric g such that the gradient vector field is "special Morse", i.e. if the gradient vector field of f in every Morse chart is the gradient of f with respect to the standard Euclidean metric on $\mathbb{R}^m$, then the unstable manifolds of (f, g) determine a CW-structure on M. A similar result was announced by Burghelea, Friedlander, and Kappeler in the Epilogue to [19] for gradient-like vector fields and proved by Audin and Damian for gradient-like vector fields in Appendix 4.9 of [4]. Qin also proved that the unstable manifolds of a Morse-Smale pair (f, g) determine a CW-structure under the assumption that the Riemannian metric is locally trivial with respect to the Morse charts [66] and then later extended the proof to general Riemannian metrics [67].

However, none of these results address the question of whether or not the CW-structure determined by the unstable manifolds is **regular**. Examples 2.14 and 2.15 and the height function on a tilted torus, cf. Example 7.11 of [8], show that the CW-structure determined by the unstable manifolds of a Morse-Smale pair (f, g) might not be regular. However, the following theorem shows that it is always possible to find at least one Morse-Smale pair (f, g) on M such that the CW-structure determined by the unstable manifolds of (f, g) is regular. In fact, it is possible to find a Morse-Smale pair (f, g) such that the unstable manifolds of (f, g) coincide with a smooth triangulation of M.

Theorem 4.12 *On any closed finite dimensional smooth manifold M there exists a smooth Morse-Smale pair (f, g) such that the unstable manifolds coincide with a smooth triangulation of M. Hence, the unstable manifolds of (f, g) determine a regular CW-structure on M. Moreover, the Riemannian metric g can be chosen such that around every critical point of f there is a Morse chart that is an isometry respect to the standard Euclidean metric on $\mathbb{R}^m$.*

Before proving this theorem, we will explain the main ideas used in the proof, give a detailed explanation of the proof for a 2-simplex, and discuss how different coordinate charts impact the construction of the Morse-Smale pair (f, g).

4.4.1 Outline of the Proof

The construction begins by choosing a smooth triangulation of M fine enough so that every closed m-cell D^m in the triangulation is contained in a coordinate chart of M, where $m < \infty$ is the dimension of M. The Morse function $f : M \to \mathbb{R}$ is then defined locally on overlapping tubular neighborhoods of the k-cells D^k for $k = 0, \ldots, m$, where the local coordinates $(x_1, \ldots, x_m)$ on the tubular neighborhoods are chosen so that $x_1, \ldots, x_k$ are coordinates on D^k, $x_{k+1}, \ldots, x_m$ are coordinates on the fibers, and both $x_1^2 + \cdots + x_k^2 < 1$ and $x_{k+1}^2 + \cdots + x_m^2 < 1$.

The function f is defined in the local coordinates using a standard form given by the Morse Lemma with a single critical point p^k of index k in the coordinate chart, e.g.

$$1 - x_1^2 - \cdots - x_k^2 + x_{k+1}^2 + \cdots + x_m^2,$$

and the standard Euclidean metric on $\mathbb{R}^m$ is pulled back to M using the local coordinate charts. This defines a Morse function and a metric such that on a neighborhood T^k of the critical point p^k we have

$$W^u(p^k)|_{T^k} = D^k|_{T^k}.$$

These locally defined functions are added together inductively for $k = 0, \ldots, m$ after multiplying by positive constants C^k and smooth bump functions $\rho : \mathbb{R} \to \mathbb{R}_+$ that are equal to 1 on $[-\frac{1}{7}, \frac{1}{7}]$ and 0 outside of $(-1, 1)$, e.g.

$$C^k \rho(x_1^2 + \cdots + x_k^2)\rho(x_{k+1}^2 + \cdots + x_m^2)(1 - x_1^2 - \cdots - x_k^2 + x_{k+1}^2 + \cdots + x_m^2).$$

This function decreases to 0 as the radial quantity $x_1^2 + \cdots + x_k^2$ in D^k goes from 0 to 1. However, it first increases and then decreases to 0 as the radial quantity $x_{k+1}^2 + \cdots + x_m^2$ in the fiber goes from 0 to 1.

This brings in "extra" critical points in the fiber of the tubular neighborhood containing p^k, in a region where coordinate neighborhoods overlap. Critical points in the regions where coordinate neighborhoods overlap are eliminated inductively for $k = 2, \ldots, m$ by choosing the constants C^k to be sufficiently large at each stage. The constants C^k are chosen large enough to ensure that the negative gradient flow of f passes through the regions where the coordinate neighborhoods overlap, towards neighborhoods of lower dimensional simplices.

4.4 A Morse-Smale Function that Determines a Regular CW-Structure

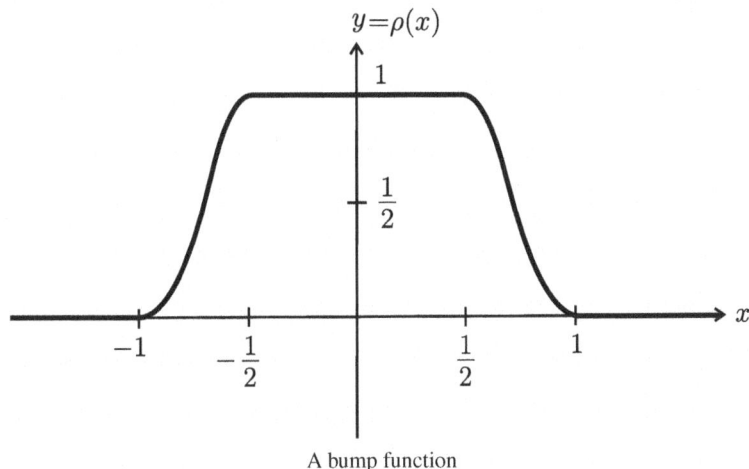

A bump function

The steps where $k = 0$ and $k = m$ are exceptional. When $k = m$ the fiber is trivial and thus there are no extra critical points, and when $k = 0$ the extra critical points are avoided by defining the function to be

$$\rho(x_1^2 + \cdots + x_m^2)(-1 + x_1^2 + \cdots + x_m^2).$$

This function increases from -1 to 0 as $x_1^2 + \cdots + x_m^2$ goes from 0 to 1.

The end result is a Morse-Smale function $f : M \to \mathbb{R}$ with a single critical point p^k of index k in the interior of each k-cell in the triangulation such that around p^k there is a neighborhood T^k with

$$W^u(p^k)|_{T^k} = D^k|_{T^k}.$$

The λ-Lemma implies that for any Morse-Smale function we have

$$\overline{W^u(p^k)} = \bigcup_{p^k \succeq p} W^u(p) \subseteq M,$$

where the notation $p^k \succeq p$ means there is a broken gradient flow line from p^k to p. (See Section 6.3 of [8].) Hence, for a sufficiently fine triangulation the construction will produce a function such that the closure of the unstable manifold $\overline{W^u(p^k)} \subseteq M$ is homeomorphic to $D^k \approx \Delta^k$. Furthermore, the construction will produce a Morse-Smale function $f : M \to \mathbb{R}$ such that $\overline{W^u(p^k)} = D^k \subseteq M$ if for every critical point $p \preceq p^k$ the coordinate chart around p is chosen so that the intersection of the j-cells in D^k with the coordinate neighborhood around p are mapped by the coordinate chart into j-dimensional affine subspaces of $\mathbb{R}^m$.

4.4.2 The Construction on a 2-Simplex

We will now illustrate the construction on a 2-simplex $M = \Delta^2 \subset \mathbb{R}^2$, where the Morse-Smale function has 3 critical points of index 0, 3 critical points of index 1, and 1 critical point of index 2. We will define local coordinates and functions around each critical point and discuss the gradient flow of the Morse-Smale function $f: M \to \mathbb{R}$ relative to the standard Euclidean metric. In order to simplify the formulas, we will use more tractable coordinate neighborhoods and basepoints here than those used in the proof of Theorem 4.12. Also, we will not give an explicit formula for the smooth bump function ρ, but we note that it satisfies the following properties:

$$\rho(x) = \begin{cases} 1 & -\tfrac{1}{2} \le x \le \tfrac{1}{2} \\ 0 & \text{if } x \notin (-1, 1), \end{cases}$$

ρ is increasing on $(-1, -\tfrac{1}{2})$, and ρ is decreasing on $(\tfrac{1}{2}, 1)$. Note that the support of ρ is $[-1, 1]$, and $\rho'(x)$ (and all its higher order derivatives) are bounded because ρ is smooth with compact support. The problems at the end of Chapter 6 in [8], Lemma 1.3 of [56], Chapter 2 of [57], and Appendix III of [95] discuss explicit formulas for smooth bump functions with the above properties.

Let $M \subset \mathbb{R}^2$ be the 2-simplex in the Euclidean (x, y)-plane with vertices

$$p_1^0 = (0, 0)$$
$$p_2^0 = (6, 0)$$
$$p_3^0 = (0, 6).$$

Around the vertex p_1^0 use the coordinates $(x_1, x_2) = (\tfrac{1}{2}x, \tfrac{1}{2}y)$ and define

$$f_1^0(x_1, x_2) = \rho(x_1^2 + x_2^2)(-1 + x_1^2 + x_2^2)$$
$$f_1^0(x, y) = \rho(\tfrac{1}{4}x^2 + \tfrac{1}{4}y^2)(-1 + \tfrac{1}{4}x^2 + \tfrac{1}{4}y^2).$$

Around the vertex p_2^0 use local coordinates $(x_1, x_2) = (\tfrac{1}{2}(x - 6), \tfrac{1}{2}y)$ and define

$$f_2^0(x_1, x_2) = \rho(x_1^2 + x_2^2)(-1 + x_1^2 + x_2^2)$$
$$f_2^0(x, y) = \rho(\tfrac{1}{4}(x - 6)^2 + \tfrac{1}{4}y^2)(-1 + \tfrac{1}{4}(x - 6)^2 + \tfrac{1}{4}y^2).$$

4.4 A Morse-Smale Function that Determines a Regular CW-Structure

Around the vertex p_3^0 use local coordinates $(x_1, x_2) = (\frac{1}{2}x, \frac{1}{2}(y-6))$ and define

$$f_3^0(x_1, x_2) = \rho(x_1^2 + x_2^2)(-1 + x_1^2 + x_2^2)$$
$$f_3^0(x, y) = \rho(\tfrac{1}{4}x^2 + \tfrac{1}{4}(y-6)^2)(-1 + \tfrac{1}{4}x^2 + \tfrac{1}{4}(y-6)^2).$$

Note that for all $j = 1, 2, 3$, the support of f_j^0 is a closed disk of radius 1 centered at p_j^0 in the local (x_1, x_2) coordinates, which is a closed disk of radius 2 in the (x, y) coordinates.

Let $p_1^1 = (3, 0)$ be a basepoint for the 1-simplex connecting p_1^0 and p_2^0, use the local coordinates $(x_1, x_2) = (\frac{1}{2}(x-3), 2y)$ around p_1^1, and define

$$f_1^1(x_1, x_2) = \rho(x_1^2)\rho(x_2^2)(1 - x_1^2 + x_2^2)$$
$$f_1^1(x, y) = \rho(\tfrac{1}{4}(x-3)^2)\rho(4y^2)(1 - \tfrac{1}{4}(x-3)^2 + 4y^2).$$

Note that the support of $f_1^1(x, y)$ is bounded by a rectangle of base 4 and height 1 centered at p_1^1, and $-(\nabla f_1^1)(x, 0)$ is parallel to the 1-simplex connecting p_1^0 and p_2^0.

Let $p_2^1 = (0, 3)$ be a basepoint for the 1-simplex connecting p_1^0 and p_3^0, use the local coordinates $(x_1, x_2) = (\frac{1}{2}(y-3), 2x)$ around p_2^1, and define

$$f_2^1(x_1, x_2) = \rho(x_1^2)\rho(x_2^2)(1 - x_1^2 + x_2^2)$$
$$f_2^1(x, y) = \rho(\tfrac{1}{4}(y-3)^2)\rho(4x^2)(1 - \tfrac{1}{4}(y-3)^2 + 4x^2).$$

Note that the support of $f_2^1(x, y)$ is bounded by a rectangle of base 1 and height 4 centered at p_2^1, and $-(\nabla f_2^1)(0, y)$ is parallel to the 1-simplex connecting p_1^0 and p_3^0.

Let $p_3^1 = (3, 3)$ be a basepoint for the 1-simplex connecting p_2^0 and p_3^0, and use the following local coordinates (x_1, x_2) around p_3^1, with x_1 along the 1-simplex and x_2 in the direction normal to the 1-simplex.

$$\begin{pmatrix} 3x_1 \\ \tfrac{1}{2}x_2 \end{pmatrix} = \begin{pmatrix} \tfrac{\sqrt{2}}{2} & -\tfrac{\sqrt{2}}{2} \\ \tfrac{\sqrt{2}}{2} & \tfrac{\sqrt{2}}{2} \end{pmatrix} \begin{pmatrix} x-3 \\ y-3 \end{pmatrix}$$

Define

$$f_3^1(x_1, x_2) = \rho(x_1^2)\rho(x_2^2)(1 - x_1^2 + x_2^2)$$
$$f_3^1(x, y) = \rho(\tfrac{1}{18}(x-y)^2)\rho(2(x+y-6)^2)$$
$$\times (73 + \tfrac{35}{18}x^2 + \tfrac{35}{18}y^2 + \tfrac{37}{9}xy - 24x - 24y).$$

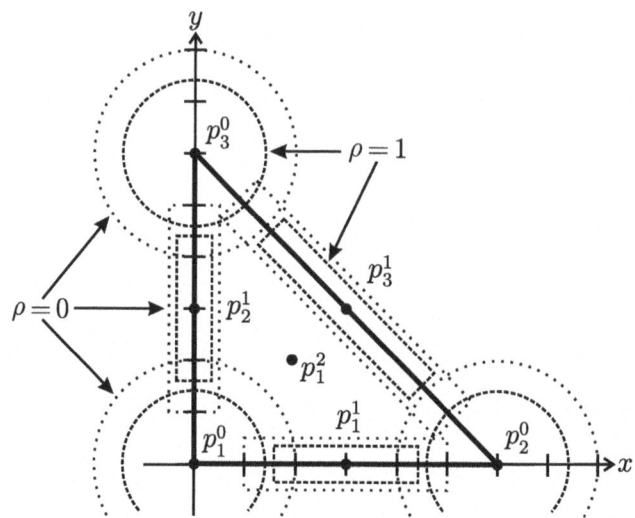

Constructing a Morse-Smale function on a 2-simplex

Note that the support of $f_3^1(x, y)$ is bounded by a rectangle of length 6 and width 1 centered on the 1-simplex containing p_3^1, and along the 1-simplex (where $y = 6-x$) we have

$$-(\nabla f_3^1)(x, 6 - x) = \tfrac{1}{9}(2x - 6)(1, -1),$$

which is parallel to the 1-simplex.

The support of f_j^1 for $j = 1, 2, 3$ is a product of two closed unit 1-disks centered at p_j^1 in the local (x_1, x_2) coordinates. In the (x, y) coordinates the support of f_1^1 is $[1, 5] \times [-\tfrac{1}{2}, \tfrac{1}{2}]$, and the support of f_2^1 is $[-\tfrac{1}{2}, \tfrac{1}{2}] \times [1, 5]$. The support of f_3^1 in the (x, y) coordinates is bounded by a rectangle centered at p_3^1 of length 6 along the 1-simplex containing p_3^1 and of width 1 in the direction normal to the 1-simplex. These choices for the local coordinates produce overlaps between the supports of the functions f_j^1 defined on tubular neighborhoods of the 1-simplices and the functions f_j^0 defined around the 0-simplices at the ends.

Define

$$f^{\leq 1} = \sum_{j=1}^{3} f_j^0 + \sum_{j=1}^{3} f_j^1,$$

and note that the support of $f^{\leq 1}$ lies in a neighborhood of the 1-skeleton of the 2-simplex M. This function has critical points of index 0 at p_j^0 and of index 1 at p_j^1 for $j = 1, 2, 3$. Since $-\nabla f_j^1$ restricted to the 1-simplex containing p_j^1 points

4.4 A Morse-Smale Function that Determines a Regular CW-Structure

in the same direction as $-\nabla f_i^0$ where the support of f_j^1 intersects the support of f_i^0 (towards the closer end of the 1-simplex), the closure of the unstable manifold $W^u(p_j^1)$ of $f^{\leq 1}$ relative to the Euclidean metric is the 1-simplex that contains p_j^1 for all $j = 1, 2, 3$.

Moreover, on the regions where the support of f_j^1 intersects the support of f_i^0, the function $f^{\leq 1}$ doesn't have any critical points because the directional derivative in the direction parallel to the 1-simplex containing p_j^1 is nonzero. This implies that any flow line of $-\nabla f^{\leq 1}$ that intersects the region where $x_2^2 < \frac{1}{2}$ in the local coordinates around a 1-simplex will flow forward in time to end at one of the critical points p_j^0 or p_j^1 for $j = 1, 2, 3$.

Note that there are 3 additional critical points inside M contained in the interior of the support of $f^{\leq 1}$. These critical points are in the fibers of the tubular neighborhoods that contain p_j^1 for $j = 1, 2, 3$. The exact location of these critical points depends on the formula for the bump function ρ, but they will lie in the region where $\frac{1}{2} \leq x_2^2 \leq 1$ in the local coordinates around a 1-simplex, i.e. where $\rho(x_2^2)$ is decreasing and $f^{\leq 1}(0, x_2) = \rho(x_2^2)(1 + x_2^2)$. Since $f^{\leq 1}$ is smooth with compact support $\|\nabla f^{\leq 1}\|$ is uniformly bounded, and the extra critical points in the interior of M can be eliminated by adding in a function f_1^2 whose gradient dominates the gradient of $f^{\leq 1}$ on the region where $-\nabla f^{\leq 1}$ points towards the interior of M.

Take $p_1^2 = (2, 2)$ as the basepoint for the 2-simplex, and use the local coordinates $(x_1, x_2) = r(x, y)(x-2, y-2)$, where $r(x, y)$ is a smooth scaling factor in the radial direction centered at p_1^2. Choose the scaling factor such that when $x_1^2 + x_2^2 = \frac{1}{2}$ in the local coordinates for the 2-simplex the point is just inside the union of the tubular neighborhoods where $x_2^2 = \frac{1}{2}$ in the local coordinates of a 1-simplex or $x_1^2 + x_2^2 = \frac{1}{2}$ in the local coordinates of a 0-simplex and when $x_1^2 + x_2^2 = 1$ in the local coordinates for the 2-simplex the point is near the union of the tubular neighborhoods where $x_2^2 = \frac{1}{4}$ in the local coordinates of a 1-simplex or $x_1^2 + x_2^2 = \frac{1}{4}$ in the local coordinates of a 0-simplex. Note that the open unit 2-ball in the local coordinates around p_1^2 contains $M - \text{supp} f^{\leq 1}$.

Define

$$f_1^2(x_1, x_2) = C_1^2 \rho(x_1^2 + x_2^2)(1 - x_1^2 - x_2^2)$$
$$f_1^2(x, y) = C_1^2 \rho(r(x, y)^2((x-2)^2 + (y-2)^2))$$
$$\times (1 - r(x, y)^2((x-2)^2 + (y-2)^2)),$$

where the constant $C_1^2 > 0$ is chosen large enough so that $\|f^{\leq 1}(x, y)\| < \|f_1^2(x, y)\|$ for all (x, y) in the compact region R where $x_1^2 + x_2^2 \leq \frac{1}{2}$ in the local coordinates around p_1^2 intersects the support of $f^{\leq 1}$, and define $f : M \to \mathbb{R}$ to be

$$f = f^{\leq 1} + f_1^2.$$

The gradient flow of the
Morse-Smale function on Δ^2

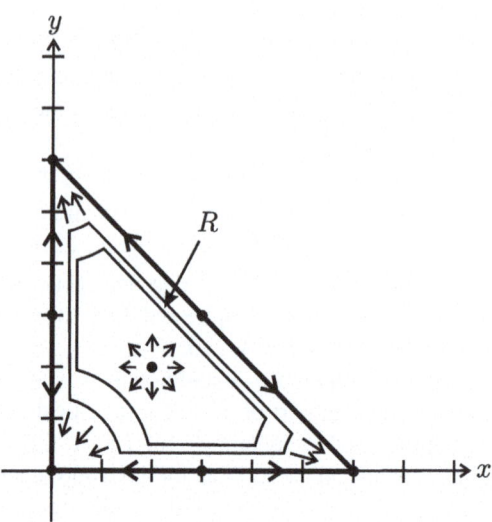

To see that the function f has no critical points where the support of $f^{\leq 1}$ intersects the support of f_1^2, note that R contains all the points where $-\nabla f^{\leq 1}$ points towards the interior of the 2-simplex and

$$\nabla f = \nabla f^{\leq 1} + \nabla f_1^2 = 0 \iff \nabla f^{\leq 1} = -\nabla f_1^2.$$

There are no solutions to this equation on the compact region R because of the choice of C_1^2, and at other points in the intersection of the supports where the two gradient vectors are parallel they point in the same direction.

Therefore, $f : M \to \mathbb{R}$ is a Morse-Smale function whose only critical points are p_j^0, p_j^1 for $j = 1, 2, 3$ and p_1^2. Moreover, the closure of $W^u(p_1^2)$ is M, and the closure of $W^u(p_j^1)$ is the 1-simplex that contains p_j^1 for all $j = 1, 2, 3$.

4.4.3 The Construction for Adjacent 2-Cells

On any smooth 2-manifold M with a sufficiently fine smooth triangulation there are smooth coordinate charts that map open neighborhoods of the 2-cells D_j^2 smoothly onto open neighborhoods of the standard 2-simplex $\Delta^2 \subset \mathbb{R}^2$, where the 0-cells and the 1-cells in D_j^2 are mapped onto the 0-faces and 1-faces of Δ^2. A Morse-Smale pair (f, g) can be defined on an open coordinate neighborhood of a 2-cell D_1^2 using a smooth coordinate chart for D_1^2, but that will also define a Morse-Smale pair on an open neighborhood of the 0-cells and 1-cells shared by an adjacent 2-cell D_2^2. Without some compatibility between the coordinate chart for D_1^2 and the 1-cells in D_2^2, the gradient flow of the sum of the gradient vector fields defined using the

4.4 A Morse-Smale Function that Determines a Regular CW-Structure

coordinate charts for D_1^2 and D_2^2 might not stay on a 1-cell where the support of the two gradient vector fields overlap, although the gradient flow will still end at a 0-cell.

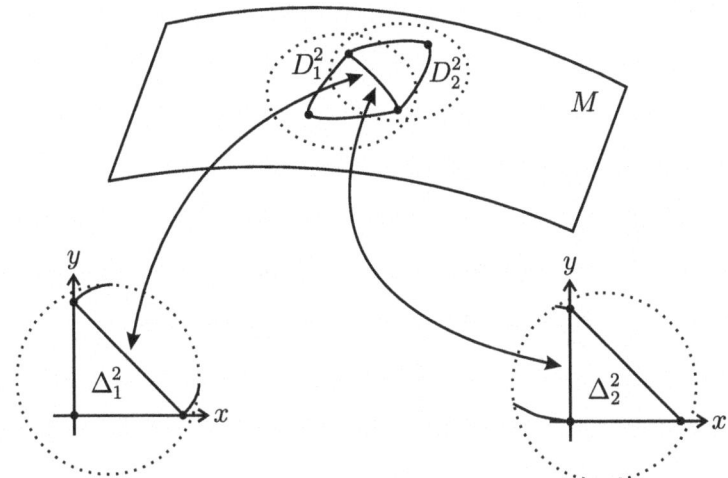

Overlapping coordinate neighborhoods for a triangulation

The concern is that, although the coordinate chart for D_2^2 will map the 1-cells in D_2^2 onto the 1-faces of Δ^2, the coordinate chart for D_1^2 might not map the intersections of the 1-cells in D_2^2 with the coordinate neighborhood of D_1^2 into affine spaces parallel to the gradient vector field defined using the coordinate chart for D_1^2. If that happens, then the sum of the gradient vector field defined around a 0-cell in D_1^2 using the coordinate chart for D_1^2 and the gradient vector field defined using the coordinate chart for D_2^2 might not be tangent to a 1-cell in the triangulation.

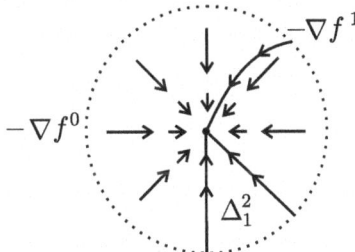

Non-parallel gradient vector fields in a neighborhood of a 0-cell

This concern can be resolved by either modifying the triangulation where it intersects the coordinate neighborhood of D_1^2 or by adjusting the coordinate chart for D_1^2 so that the intersections of the 1-cells in the triangulation with the coordinate neighborhood of D_1^2 are mapped into 1-dimensional affine subspaces of $\mathbb{R}^2$ by the coordinate chart. This will ensure that the gradient vector field defined using the coordinate chart for D_1^2 and the gradient vector fields defined on the 1-cells adjacent to D_1^2 (using different coordinate charts) will be parallel where their supports overlap. One way to define such a change of coordinates is to use the Isotopy (Diffeotopy) Extension Theorem, cf. Theorem 8.1.3 of [39], Theorem 7.3.3 and Corollary 7.3.4 of [55], or Theorem 2.4.2 of [88].

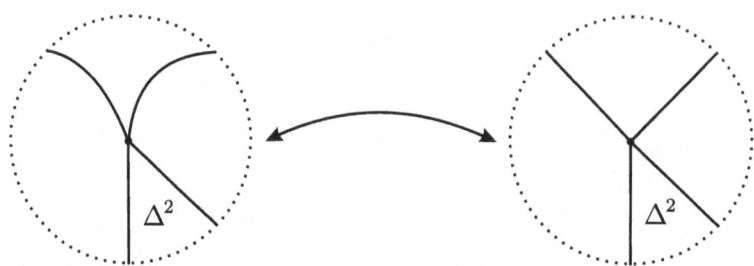

Changing coordinates around a 0-cell to make the adjacent 1-cells affine

More specifically, consider a 1-cell D^1 in a 2-cell adjacent to D_1^2 that has a boundary point D^0 in D_1^2. The closure of the image of D^1 under the coordinate chart for D_1^2 is isotopic to a compact set A contained in the 1-dimensional affine subspace of $\mathbb{R}^2$ that contains the image p^0 of D^0 and the other boundary point q^0 of the image of D^1 restricted to the coordinate neighborhood. Assuming that the coordinate neighborhood of D_1^2 doesn't extend too far beyond D_1^2, the image of the isotopy on the interior of the image of D^1 will be contained in an isolating neighborhood $U \subset \mathbb{R}^2$, i.e. an open set that does not intersect the image of another 1-cell. The Isotopy (Diffeotopy) Extension Theorem implies that there is a 1-parameter family of diffeomorphisms φ_t with support in $\overline{U}$ that extends the isotopy, and the diffeomorphism φ_1 maps the image of the intersection of D^1 with the coordinate neighborhood onto $A - q^0$. Note that the endpoints p^0 and q^0 are not moved by the isotopy, and thus they are fixed points of φ_1.

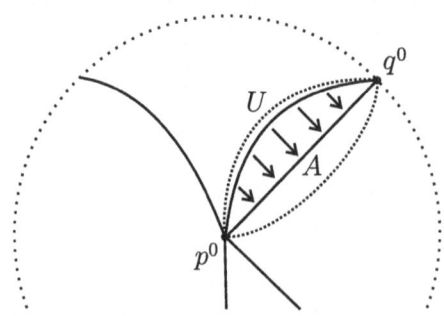

A diffeomorphism induced by an isotopy

Repeating this argument for each adjacent 1-cell that intersects the coordinate neighborhood of D_1^2 gives a finite number of diffeomorphisms on the image of the coordinate chart in $\mathbb{R}^2$. Composing these diffeomorphisms with the original coordinate chart for D_1^2 gives a coordinate chart where the intersections of the adjacent 1-cells in the triangulation with the coordinate neighborhood of D_1^2 are mapped into 1-dimensional affine subspaces of $\mathbb{R}^2$.

Remark 4.13 (The Isotopy (Diffeotopy) Extension Theorem) A smooth isotopy $h : V \times [0, 1] \to U$ of a compact smooth manifold V (possibly with boundary) to a smooth manifold U (possibly with boundary) determines a level preserving map $\hat{h} : V \times [0, 1] \to U \times [0, 1]$, given by $\hat{h}(x, t) = (h(x, t), t)$. Points $(x, t) \in V \times [0, 1]$ determine arcs $\hat{h}(\{x\} \times [0, 1])$ in $U \times [0, 1]$, whose tangent vectors determine a smooth vector field X on the image $\hat{h}(V \times [0, 1]) \subseteq U \times [0, 1]$. The Isotopy (Diffeotopy) Extension Theorem is proved by observing that X can be extended to a smooth vector field on $U \times [0, 1]$ with compact support whenever $h((\partial V) \times [0, 1]) \subseteq \partial U$ or $h(V \times [0, 1]) \cap \partial U = \emptyset$. Thus, X determines a diffeotopy $\varphi : U \times [0, 1] \to U$ with compact support that extends the isotopy. For more details see Theorem 8.1.3 of [39], Theorem 7.3.3 and Corollary 7.3.4 of [55], or Theorem 2.4.2 of [88].

In the proof of Theorem 4.12, we start with a sufficiently fine smooth triangulation and then choose coordinate charts using coordinate transformations such as those coming from the Isotopy (Diffeotopy) Extension Theorem to ensure that the intersection of the k-cells in the triangulation with various tubular neighborhoods are mapped into k-dimensional affine subspace of $\mathbb{R}^m$ in the local coordinates. For a description of a more specific triangulation that works well for the proof of Theorem 4.12 see Remark 4.14.

4.4.4 Proof of Theorem 4.12

Following an idea dual to a method used in one of Thurston's unpublished manuscripts, Chapter III of [6], we choose a smooth triangulation of M fine enough so that every m-cell $D^m \subset M$ in the triangulation has a coordinate chart on a neighborhood of D^m that sends D^m to the standard m-simplex $\Delta^m \subset \mathbb{R}^m$, where $m < \infty$ is the dimension of the manifold. Note that this implies that every k-cell $D^k \subset D^m$, for $k = 0, \ldots, m-1$, inherits coordinates from the coordinate chart sending D^m to Δ^m, and the k-cells contained in D^m are mapped to the k-faces of Δ^m under the coordinate chart. Let n_k denote the number of k-cells in the triangulation for all $k = 0, \ldots, m$, let D_j^k denote the jth k-cell for all $j = 1, \ldots, n_k$, let $p_j^0 = D_j^0$ for all $j = 1, \ldots, n_0$, and for each $k > 0$ pick a basepoint $p_j^k \in D_j^k$ for all $j = 1, \ldots, n_k$ near the barycenter with respect to coordinates determined by a chart that sends an m-cell D^m containing D_j^k to $\Delta^m \subset \mathbb{R}^m$.

Pick a smooth Riemannian metric $\tilde{g}$ on M, and let $\rho : \mathbb{R} \to \mathbb{R}_+$ be a smooth bump function such that

- $\rho(-x) = \rho(x)$ for all x,
- $0 < \rho(x) \leq 1$ if $|x| < 1$,
- $\rho(x) = 0$ if $|x| \geq 1$,
- $\rho(x) = 1$ if $|x| \leq \frac{1}{2}$,
- $\rho'(x) > 0$ if $-1 < x < -\frac{1}{2}$ and $\rho'(x) < 0$ if $\frac{1}{2} < x < 1$,
- $\rho^{(n)}(\pm 1) = 0$ for all $n \geq 0$.

(See the problems at the end of Chapter 6 in [8], Lemma 1.3 of [56], Chapter 2 of [57], or Lemma 1b in Appendix III of [95].)

Around every 0-cell $D_j^0 = p_j^0$ choose a small open coordinate neighborhood $T_{j,1}^0$ centered at p_j^0 such that $\overline{T_{i,1}^0} \cap \overline{T_{j,1}^0} = \emptyset$ if $i \neq j$ and p_j^0 is the only basepoint in $\overline{T_{j,1}^0}$. By making a change of coordinates on $T_{j,1}^0$ we will assume that in the local coordinates $(x_1, \ldots, x_m)$ we have

$$T_{j,1}^0 = \{(x_1, \ldots, x_m) | x_1^2 + \cdots + x_m^2 < 1\},$$

where $(0, \ldots, 0)$ corresponds to the point p_j^0, and the intersection of every k-cell in the triangulation with $T_{j,1}^0$ is mapped into a k-dimensional affine subspace of $\mathbb{R}^m$ in the local coordinates $(x_1, \ldots, x_m)$, cf. [95]. Inside each coordinate neighborhood we have smaller coordinate neighborhoods $T_{j,s}^0 \subseteq T_{j,1}^0$, where $s \in (0, 1]$.

$$T_{j,s}^0 = \left\{(x_1, \ldots, x_m) \in T_{j,1}^0 \,\Big|\, x_1^2 + \cdots + x_m^2 < s\right\}$$

Modify $\tilde{g}$ on $T_{j,\frac{1}{2}}^0$ to a smooth metric g^0 that is the pullback of the standard metric on $\mathbb{R}^m$ under the coordinate chart, cf. Remark 6.31 of [8], and define a function

$$f_j^0(x_1, \ldots, x_m) = \rho(x_1^2 + \cdots + x_m^2)(-1 + x_1^2 + \cdots + x_m^2)$$

using the local coordinates on $T_{j,1}^0$. Note that

$$f_j^0(x_1, \ldots, x_m) = -1 + x_1^2 + \cdots + x_m^2 \text{ for all } (x_1, \ldots, x_m) \in T_{j,\frac{1}{2}}^0$$

and for all $(x_1, \ldots, x_m) \in T_{j,1}^0$ we have

$$\frac{\partial f_j^0}{\partial x_i}(x_1, \ldots, x_m) < 0 \text{ if } x_i < 0$$

$$\frac{\partial f_j^0}{\partial x_i}(x_1, \ldots, x_m) > 0 \text{ if } x_i > 0$$

4.4 A Morse-Smale Function that Determines a Regular CW-Structure

for all $i = 1, \ldots, m$. Moreover, all the higher order partial derivatives of f_j^0 are zero on the boundary of $T_{j,1}^0$, and hence we can extend f_j^0 to a smooth function on M by setting it equal to zero outside of $T_{j,1}^0$. For all $s \in (0, 1]$ define

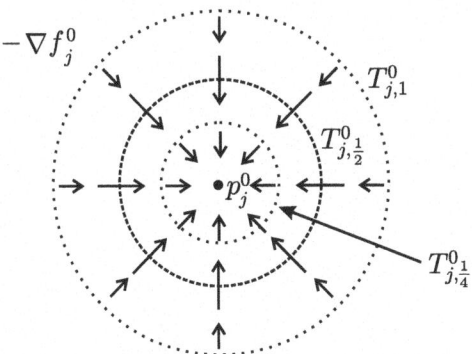

$$T_s^0 = \bigcup_{j=1}^{n_0} T_{j,s}^0 \quad \text{and} \quad f^0 = \sum_{j=1}^{n_0} f_j^0.$$

The function $f^0 : M \to \mathbb{R}$ is a smooth function that is Morse-Smale with respect to the metric $\mathbf{g}^0$ on $T_{\frac{1}{2}}^0$, it has a critical point p_j^0 of index 0 at each 0-cell D_j^0 in the triangulation, and it is equal to zero on $M - T_1^0$.

Next, for every 1-cell D_j^1 in the triangulation choose a small open tubular neighborhood $T_{j,1,1}^1$ of $D_j^1 - \overline{T_{\frac{1}{4}}^0}$ with coordinates $(x_1, \ldots, x_m)$ centered at the basepoint $p_j^1 \in D_j^1$, where x_1 is the coordinate along the cell D_j^1 and $(x_2, \ldots, x_m)$ are coordinates normal to D_j^1. Choose the tubular neighborhoods small enough so that $\overline{T_{i,1,1}^1} \cap \overline{T_{j,1,1}^1} = \emptyset$ if $i \neq j$ and p_j^1 is the only basepoint in $\overline{T_{j,1,1}^1}$. By making a change of coordinates on $T_{j,1,1}^1$ we will assume that in the local coordinates $(x_1, \ldots, x_m)$ we have

$$T_{j,1,1}^1 = \{(x_1, \ldots, x_m) | \, x_1^2 < 1 \text{ and } x_2^2 + \cdots + x_m^2 < 1\},$$

where $(0, \ldots, 0)$ corresponds to p_j^1, and the intersection of every k-cell in the triangulation with $T_{j,1,1}^1$ is mapped into a k-dimensional affine subspace of $\mathbb{R}^m$ in the local coordinates $(x_1, \ldots, x_m)$. Inside each coordinate neighborhood we have smaller open tubular neighborhoods $T_{j,r,s}^1 \subseteq T_{j,1,1}^1$, where $r, s \in (0, 1]$.

$$T_{j,r,s}^1 = \left\{ (x_1, \ldots, x_m) \in T_{j,1,1}^1 \,\Big|\, x_1^2 < r \text{ and } x_2^2 + \cdots + x_m^2 < s \right\}$$

By making another change of coordinates on $T^1_{j,1,1}$ we will assume that the boundary of the 1-disk

$$D^1_{j,\frac{1}{2}}(x_1) = \left\{(x_1,\ldots,x_m) \in T^1_{j,1,1} \,\Big|\, x_1^2 < \frac{1}{2} \text{ and } x_2 = \cdots = x_m = 0\right\},$$

i.e. $(\pm\frac{1}{\sqrt{2}}, 0, \ldots, 0)$, is contained in $T^0_{\frac{1}{2}}$, the fibers of $T^1_{j,\frac{1}{2},\frac{1}{2}}$ above $(\pm\frac{1}{\sqrt{2}}, 0, \ldots, 0)$ are contained in $T^0_{\frac{1}{2}}$, and increasing the radial quantity x_1^2 in the coordinate system for $T^1_{j,1,1}$ decreases the radial quantity $x_1^2 + \cdots + x_m^2$ in the coordinates for T^0_1 on $T^1_{j,1,1} \cap T^0_1$.

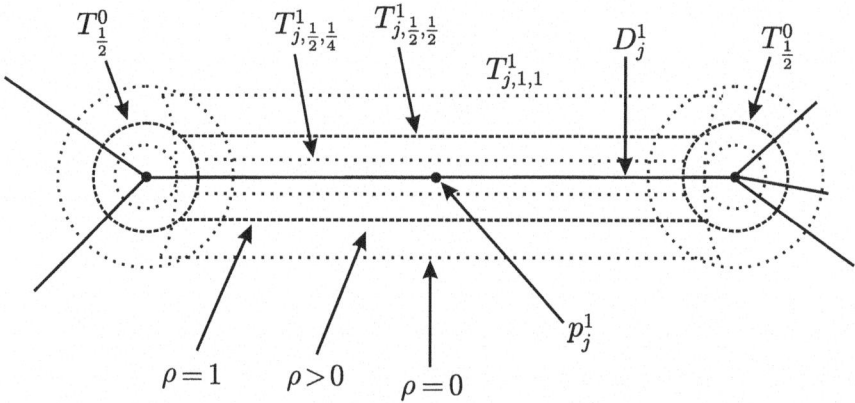

Modify the metric g^0 to a smooth metric g^1 that is the pullback of the standard metric on $\mathbb{R}^m$ under the coordinate chart on $T^1_{j,\frac{1}{2},\frac{1}{2}}$ and define a function

$$f^1_j(x_1,\ldots,x_m) = \rho(x_1^2)\rho(x_2^2 + \cdots + x_m^2)(1 - x_1^2 + x_2^2 + \cdots + x_m^2)$$

using the local coordinates on $T^1_{j,1,1}$. We have

$$f^1_j(x_1,\ldots,x_m) = 1 - x_1^2 + x_2^2 + \cdots + x_m^2 \text{ for all } (x_1,\ldots,x_m) \in T^1_{j,\frac{1}{2},\frac{1}{2}},$$

and all the higher order partial derivatives of f^1_j are zero on the boundary of $T^1_{j,1,1}$. Hence, we can extend f^1_j to a smooth function on M by setting it equal to 0 outside of $T^1_{j,1,1}$. Note that for all $(x_1,\ldots,x_m) \in T^1_{j,1,1}$,

$$\frac{\partial f^1_j}{\partial x_1}(x_1,\ldots,x_m) > 0 \text{ if } x_1 < 0$$

$$\frac{\partial f^1_j}{\partial x_1}(x_1,\ldots,x_m) < 0 \text{ if } x_1 > 0.$$

4.4 A Morse-Smale Function that Determines a Regular CW-Structure

Thus, the critical points of f_j^1 are all contained in the fiber of the tubular neighborhood containing p_j^1, i.e. where $x_1 = 0$.

For all $r, s \in (0, 1]$ define

$$T_{r,s}^1 = \bigcup_{j=1}^{n_1} T_{j,r,s}^1, \quad T_{r,s}^{\leq 1} = T_s^0 \cup T_{r,s}^1, \quad f^1 = \sum_{j=1}^{n_1} f_j^1 \quad \text{and} \quad f^{\leq 1} = f^0 + f^1.$$

The function $f^{\leq 1}$ is a smooth function on M that is Morse-Smale with respect to the metric g^1 on $T_{\frac{1}{2},\frac{1}{2}}^{\leq 1}$. It has a critical point p_j^0 of index 0 at each 0-cell D_j^0 in the triangulation, and the closure of the unstable manifold $W^u(p_j^1)$ is $D_j^1 \approx \Delta^1$ for each basepoint $p_j^1 \in D_k^1$. Moreover, there are Morse charts for $f^{\leq 1}$ that are isometries with respect to g^1 and the standard Euclidean metric on $\mathbb{R}^m$ around p_j^k for $k = 0, 1$ and $j = 1, \ldots, n_k$.

For all $j = 1, \ldots, n_1$, the only critical point of $f^{\leq 1}$ in $T_{j,1,\frac{1}{2}}^1 - T_1^0$ is p_j^1, since $f^0 \equiv 0$ and $\rho(x_2^2 + \cdots + x_m^2) = 1$ on that region. We claim that $f^{\leq 1}$ has no critical points on the region $T_{j,1,1}^1 \cap T_1^0$, and hence p_j^1 is the only critical point of $f^{\leq 1}$ in $T_{j,1,\frac{1}{2}}^1$. To see this, recall that f_j^1 decreases as x_1^2 increases, and note that the coordinate charts on T_1^0 and $T_{j,1,1}^1$ were chosen so that f^0 also decreases as x_1^2 in the coordinate system of $T_{j,1,1}^1$ increases. Thus, $f^{\leq 1} = f^0 + f^1$ has a nonzero directional derivative at every point in $T_{j,1,1}^1 \cap T_1^0$. Note that this implies that any flow line of $-\nabla f^{\leq 1}$ that intersects the region $T_{\frac{1}{2},\frac{1}{2}}^{\leq 1}$ will flow forward in time to the basepoint of a 1-cell or a 0-cell.

However, the bump function produces additional critical points in the fiber containing p_j^1, where in the coordinates for $T_{j,1,1}^1$ we have

$$f^{\leq 1}|_{T_{j,1,1}^1}(0, x_2, \ldots, x_m) = f_j^1(0, x_2, \ldots, x_m) = \rho(x_2^2 + \cdots + x_m^2)(1 + x_2^2 + \cdots + x_m^2).$$

We will eliminate these critical points by adding in functions f^j for $j = 2, \ldots, m$ whose gradients dominate the gradient of $f^{\leq 1}$. This is possible because $f^{\leq 1}$ is smooth with compact support, which implies that the magnitude of the gradient of $f^{\leq 1}$ is uniformly bounded.

For every 2-cell D_j^2 in the triangulation choose a small open tubular neighborhood $T_{j,1,1}^2$ of

$$D_j^2 - \overline{T_{1,\frac{1}{4}}^{\leq 1}}$$

with coordinates $(x_1, \ldots, x_m)$ centered at the basepoint point $p_j^2 \in D_j^2$, where (x_1, x_2) are coordinates on the cell D_j^2 and $(x_3, \ldots, x_m)$ are coordinates normal to D_j^2. Choose the tubular neighborhoods small enough so that $\overline{T_{i,1,1}^2} \cap \overline{T_{j,1,1}^2} = \emptyset$ if $i \neq j$ and p_j^2 is the only basepoint in $\overline{T_{j,1,1}^2}$. By making a change of coordinates on $T_{j,1,1}^2$ we will assume that in the local coordinates $(x_1, \ldots, x_m)$ we have

$$T_{j,1,1}^2 = \{(x_1, \ldots, x_m) | x_1^2 + x_2^2 < 1 \text{ and } x_3^2 + \cdots + x_m^2 < 1\},$$

where $(0, \ldots, 0)$ corresponds to p_j^2, and the intersection of every k-cell in the triangulation with $T_{j,1,1}^2$ is mapped into a k-dimensional affine subspace of $\mathbb{R}^m$ in the local coordinates $(x_1, \ldots, x_m)$. Inside each coordinate neighborhood we have smaller open tubular neighborhoods $T_{j,r,s}^2 \subseteq T_{j,1,1}^2$ and 2-disks $D_{j,r}^2(x_1, x_2)$, where $r, s \in (0, 1]$.

$$T_{j,r,s}^2 = \left\{(x_1, \ldots, x_m) \in T_{j,1,1}^2 \,\Big|\, x_1^2 + x_2^2 < r \text{ and } x_3^2 + \cdots + x_m^2 < s\right\}$$
$$D_{j,r}^2(x_1, x_2) = \left\{(x_1, \ldots, x_m) \in T_{j,1,1}^2 \,\Big|\, x_1^2 + x_2^2 < r \text{ and } x_3 = \cdots = x_m = 0\right\}$$

By making another change of coordinates on $T_{j,1,1}^2$ we will assume that the boundary of the 2-disk $D_{j,\frac{1}{2}}^2(x_1, x_2)$ is contained in $T_{1,\frac{1}{2}}^{\leq 1}$, the fibers of $T_{j,\frac{1}{2},\frac{1}{2}}^2$ above the boundary points of $D_{j,\frac{1}{2}}^2(x_1, x_2)$ are contained in $T_{1,\frac{1}{2}}^{\leq 1}$, and increasing the radial quantity $x_1^2 + x_2^2$ in the coordinate system for $T_{j,1,1}^2$ decreases the radial quantity $x_1^2 + \cdots + x_m^2$ in the coordinates for T_1^0 on $T_{j,1,1}^2 \cap T_1^0$ and the radial quantity $x_2^2 + \cdots + x_m^2$ in the coordinates for $T_{1,1}^1$ on $T_{j,1,1}^2 \cap T_{1,1}^1$.

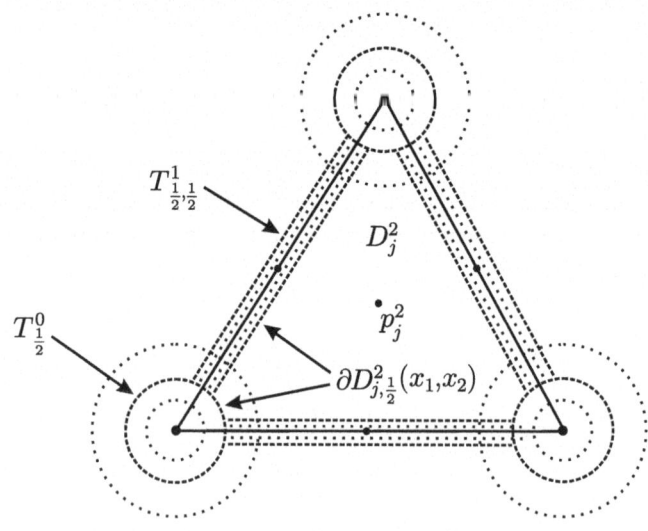

4.4 A Morse-Smale Function that Determines a Regular CW-Structure

Modify the metric g^1 to a smooth metric g^2 that is the pullback of the standard metric on $\mathbb{R}^m$ under the coordinate chart on $T^2_{j,\frac{1}{2},\frac{1}{2}}$ and define a function

$$f_j^2(x_1,\ldots,x_m) = C_j^2 \, \rho(x_1^2+x_2^2)\rho(x_3^2+\cdots+x_m^2)(1-x_1^2-x_2^2+x_3^2+\cdots+x_m^2)$$

using the local coordinates on $T^2_{j,1,1}$, where $C_j^2 > 0$ is chosen large enough so that $f^{\leq 1} + f_j^2$ does not have any critical points in $T^{\leq 1}_{1,1} \cap T^2_{j,1,\frac{1}{2}}$. The constant C_j^2 exists because f_j^2 decreases as the radial quantity $x_1^2+x_2^2$ increases, and there is a uniform bound on the magnitude of the gradient of $f^{\leq 1} : M \to \mathbb{R}$ (with respect to the metric g^2) on the region where the gradient of $f^{\leq 1}$ could point in the opposite direction of the gradient of f_j^2, i.e.

$$\left(T^{\leq 1}_{1,1} - \overline{T^{\leq 1}_{1,\frac{1}{2}}}\right) \cap T^2_{j,1,\frac{1}{2}}.$$

We have

$$f_j^2(x_1,\ldots,x_m) = C_j^2\,(1-x_1^2-x_2^2+x_3^2+\cdots+x_m^2) \text{ for all } (x_1,\ldots,x_m) \in T^2_{j,\frac{1}{2},\frac{1}{2}},$$

and all the higher order partial derivatives of f_j^2 are zero on the boundary of $T^2_{j,1,1}$. Hence, we can extend f_j^2 to a smooth function on M by setting it equal to 0 outside of $T^2_{j,1,1}$. The critical points of f_j^2 are all contained in the fiber of the tubular neighborhood containing p_j^2, i.e. where $x_1 = x_2 = 0$, because for all $(x_1,\ldots,x_m) \in T^2_{j,1,1}$ we have

$$\frac{\partial f_j^2}{\partial x_i}(x_1,\ldots,x_m) > 0 \text{ if } x_i < 0$$

$$\frac{\partial f_j^2}{\partial x_i}(x_1,\ldots,x_m) < 0 \text{ if } x_i > 0$$

for $i = 1, 2$.

For all $r, s \in (0, 1]$ define

$$T^2_{r,s} = \bigcup_{j=1}^{n_2} T^2_{j,r,s}, \quad T^{\leq 2}_{r,s} = T^{\leq 1}_{r,s} \cup T^2_{r,s}, \quad f^2 = \sum_{j=1}^{n_2} f_j^2 \quad \text{and} \quad f^{\leq 2} = f^{\leq 1} + f^2.$$

The function $f^{\leq 2}$ is a smooth function on M that is Morse-Smale with respect to the metric g^2 on $T^{\leq 2}_{\frac{1}{2},\frac{1}{2}}$. Its critical points consist of the basepoints $p_j^k \in D_j^k$ for every k-cell in the triangulation for $k = 0, 1, 2$ and some critical points in the fibers containing the critical points p_j^2, where we have

$$f^{\leq 2}|_{T_{j,1,1}^2}(0, 0, x_3, \ldots, x_m) = f_j^2(0, 0, x_3, \ldots, x_m)$$
$$= C_j^2 \rho(x_3^2 + \cdots + x_m^2)(1 + x_3^2 + \cdots + x_m^2)$$

in the local coordinates for $T_{j,1,1}^2$. We will eliminate these critical points by adding in functions f^j for $j = 3, \ldots, m$ whose gradients dominate the gradient of $f^{\leq 2}$ on these fibers.

For the pair $(f^{\leq 2}, \mathbf{g}^2)$, the closure of the unstable manifold $W^u(p_j^1)$ is $D_j^1 \approx \Delta^1$ for each critical point p_j^1, and the closure of the unstable manifold $W^u(p_j^2)$ is $D_j^2 \approx \Delta^2$ for each critical point p_j^2. Moreover, there are Morse charts for $f^{\leq 2}$ that are isometries with respect to $\mathbf{g}^2$ and the standard Euclidean metric on $\mathbb{R}^m$ around every critical point in $T_{\frac{1}{2},\frac{1}{2}}^{\leq 2}$, and any flow line of $-\nabla f^{\leq 2}$ that intersects $T_{\frac{1}{2},\frac{1}{2}}^{\leq 2}$ will flow forward in time to the basepoint of an i-cell where $i \leq 2$.

Continuing in this fashion, there are open tubular neighborhoods

$$T_{r,s}^{\leq k} = T_s^0 \cup T_{r,s}^1 \cup \cdots \cup T_{r,s}^k,$$

where $r, s \in (0, 1]$ and a pair $(f^{\leq k}, \mathbf{g}^k)$ that satisfy the following for all $k = 0, \ldots, m-1$ and $j = 1, \ldots, n_k$.

- The function $f^{\leq k} : M \to \mathbb{R}$ is smooth and Morse-Smale with respect to $\mathbf{g}^k$ on $T_{\frac{1}{2},\frac{1}{2}}^{\leq k}$.
- The basepoints $p_j^k \in D_j^k$ are the only critical points of $f^{\leq k}$ in $T_{1,\frac{1}{2}}^{\leq k}$.
- There is a Morse chart for $f^{\leq k}$ around each critical point p_j^k that is an isometry with respect to $\mathbf{g}^k$ and the standard Euclidean metric on $\mathbb{R}^m$.
- The closure of the unstable manifold $W^u(p_j^k)$ is the cell $D_j^k \approx \Delta^k$.
- Every flow line of $-\nabla f^{\leq k}$ that intersects $T_{\frac{1}{2},\frac{1}{2}}^{\leq k}$ will flow forward in time to the basepoint of an i-cell, where $i \leq k$.

Finally, around every critical point p_j^m we have the open neighborhood

$$T_{j,1}^m = D_j^m - \overline{T_{1,\frac{1}{4}}^{\leq m-1}}$$

with coordinates $(x_1, \ldots, x_m)$ centered at the basepoint $p_j^m \in D_j^m$. By making a change of coordinates on $T_{j,1}^m$ we will assume that in the local coordinates we have

$$T_{j,1}^m = \{(x_1, \ldots, x_m) \mid x_1^2 + \cdots + x_m^2 < 1\},$$

4.4 A Morse-Smale Function that Determines a Regular CW-Structure

where $(0, \ldots, 0)$ corresponds to the point p_j^m. Inside each coordinate neighborhood we have smaller coordinate neighborhoods $T_{j,r}^m \subseteq T_{j,1}^m$, where $r \in (0, 1]$.

$$T_{j,r}^m = \left\{ (x_1, \ldots, x_m) \in T_{j,1}^m \;\middle|\; x_1^2 + \cdots + x_m^2 < r \right\}$$

By making another change of coordinates on $T_{j,1}^m$ we will assume the boundary of $T_{j,\frac{1}{2}}^m$ is contained in $T_{1,\frac{1}{2}}^{\leq m-1}$ and increasing the radial quantity $x_1^2 + \cdots + x_m^2$ in the coordinate system for $T_{j,1}^m$ decreases the radial quantity $x_1^2 + \cdots + x_m^2$ in the coordinates for T_1^0 on $T_{j,1}^m \cap T_1^0$ and the radial quantity $x_k^2 + \cdots + x_m^2$ in the coordinates for $T_{1,1}^k$ on $T_{j,1}^m \cap T_{1,1}^k$ for all $k = 1, \ldots, m-1$.

Modify g^{m-1} on $T_{j,\frac{1}{2}}^m$ to a smooth metric $\mathsf{g}^m = \mathsf{g}$ that is the pullback of the standard metric on $\mathbb{R}^m$ under the coordinate chart, and define a function

$$f_j^m(x_1, \ldots, x_m) = C_j^m \rho(x_1^2 + \cdots + x_m^2)(1 - x_1^2 - \cdots - x_m^2)$$

using the local coordinates on $T_{j,1}^m$, where C_j^m is chosen large enough so that $f^{\leq m-1} + f_j^m$ does not have any critical points in $T_{j,1}^m - \{p_j^m\}$. Note that

$$f_j^m(x_1, \ldots, x_m) = C_j^m (1 - x_1^2 - \cdots - x_m^2) \text{ for all } (x_1, \ldots, x_m) \in T_{j,\frac{1}{2}}^m,$$

and for $(x_1, \ldots, x_m) \in T_{j,1}^m$ we have

$$\frac{\partial f_j^m}{\partial x_i}(x_1, \ldots, x_m) > 0 \text{ if } x_i < 0$$

$$\frac{\partial f_j^m}{\partial x_i}(x_1, \ldots, x_m) < 0 \text{ if } x_i > 0$$

for all $i = 1, \ldots, m$. Moreover, all the higher order partial derivatives of f_j^m are zero on the boundary of $T_{j,1}^m$, and hence we can extend f_j^m to a smooth function on M by setting it equal to zero outside of $T_{j,1}^m$. Define

$$f^m = \sum_{j=1}^{n_m} f_j^m \quad \text{and} \quad f = f^{\leq m} = f^{\leq m-1} + f^m.$$

The pair (f, g) is a smooth Morse-Smale pair on M that satisfies the following for all $k = 0, \ldots, m$ and $j = 1, \ldots, n_k$.

- The basepoints p_j^k are the critical points of f of index k.
- There is a Morse chart for f around each critical point p_j^k that is an isometry with respect to g and the standard Euclidean metric on $\mathbb{R}^m$.

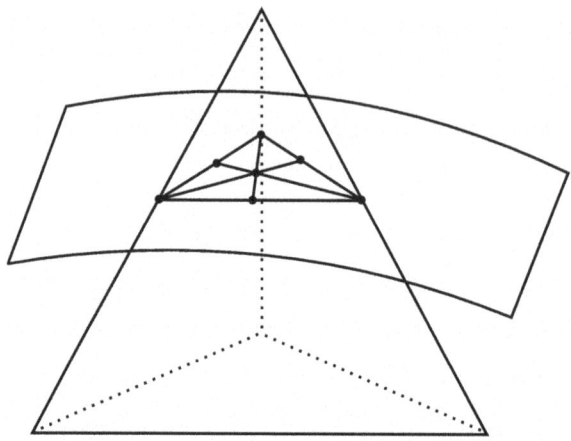

Whitney's triangulation for a surface embedded in $\mathbb{R}^3$

- The closure of the unstable manifold $W^u(p_j^k)$ of the critical point p_j^k is the cell $D_j^k \approx \Delta^k$.

Therefore, the unstable manifolds of (f, g) coincide with the smooth triangulation, and there is a Morse chart around every critical point of f that is an isometry with respect to the standard Euclidean metric on $\mathbb{R}^m$.

This completes the proof of Theorem 4.12. □

Remark 4.14 (Triangulations) It is well known that any finite dimensional compact smooth manifold has a smooth triangulation [20, 21, 56, 94, 95]. However, some triangulations are more useful than others from a computational point of view. For instance, the modification of Whitney's construction in [14] gives a computationally efficient triangulation that works well for the Morse-Smale function constructed in Theorem 4.12.

Whitney's construction begins by embedding a smooth manifold M of dimension $m < \infty$ into $\mathbb{R}^n$ for some $n > m$, and then it proceeds by finding a triangulation of $\mathbb{R}^n$ that is in general position relative to $M \subset \mathbb{R}^n$. That is, M does not intersect the simplices in the triangulation of $\mathbb{R}^n$ of dimension strictly less than $n - m$, and it is transverse to the simplices of dimension greater than or equal to $n - m$. In [14] this is achieved by starting with a Coxeter triangulation of type $\tilde{A}_n$ of $\mathbb{R}^n$, and then the vertices of the triangulation are perturbed so that the simplices in the triangulation are all transverse to M. Alternately, one could start with a triangulation of $\mathbb{R}^n$ and perturb the embedding of M to make it transverse to the simplices in the triangulation.

The triangulation of $M \subset \mathbb{R}^n$ is created from the triangulation of $\mathbb{R}^n$ by assigning a unique point $v(\tau)$ to each simplex τ in the triangulation of $\mathbb{R}^n$ that intersects M. When M intersects a simplex τ of dimension $n - m$, the point $v(\tau)$ is unique, and for a higher dimensional simplex τ the point $v(\tau)$ is chosen as the barycenter of the points where M intersects the faces of τ of dimension $n - m$. The simplices in the triangulation of M are then defined by $\sigma^k = \{v(\tau_0), \ldots, v(\tau_k)\}$ for every sequence

4.4 A Morse-Smale Function that Determines a Regular CW-Structure

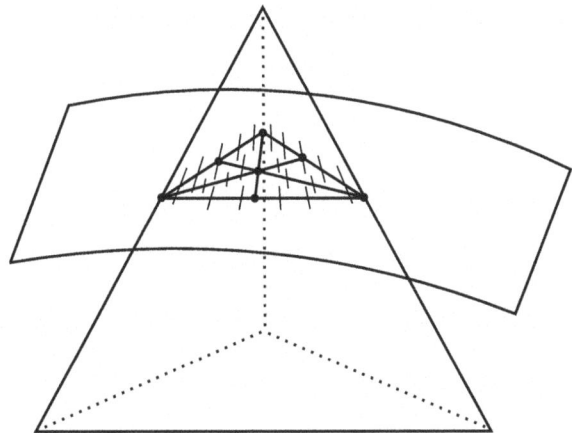

Fibers in the tubular neighborhood of K [14]

$\tau_0 \subset \tau_1 \subset \cdots \subset \tau_k$ of distinct simplices in the triangulation of $\mathbb{R}^n$ such that τ_0 intersects M, cf. Section 6.2 of [14] or Section IV.20 of [95].

The union of the simplices σ^k form a simplicial complex $K \subset \mathbb{R}^n$ that is homeomorphic to $M \subset \mathbb{R}^n$. In fact, there is a piecewise smooth diffeomorphism $M \to K$ defined by projecting along the fibers of a piecewise smooth tubular neighborhood of K that contains M. The fibers of the tubular neighborhood are affine subspaces of $\mathbb{R}^n$ of dimension $n-m$ that contain the simplices τ of dimension $n-m$ in the triangulation of $\mathbb{R}^n$ whenever $\tau \cap M \neq \emptyset$, see Section 7 of [14]. Smoothing the corners of the projection maps give local coordinate charts which map the intersection of adjacent j-cells into j-dimensional affine subspaces.

Remark 4.15 (Combinatorial Morse Theory) The smooth Morse-Smale pair (f, g) in Theorem 4.12 was constructed by starting with a sufficiently fine triangulation of M and then defining f and g so that the closure of the unstable manifolds of (f, g) coincide with the chosen triangulation. Hence, it seems likely that Theorem 4.12 will have applications to combinatorial Morse theory. For instance, Theorem 3.1 of [33] proves that given a generic Morse function $f : M \to \mathbb{R}$ on a smooth closed oriented manifold M there exists a C^1-triangulation T of M and a combinatorial Morse vector field V on T that realizes the smooth Morse-Smale-Witten chain complex. Theorem 4.12 should be useful for proving a converse to this theorem.

Remark 4.16 (The Fundamental Identity and Theorem 4.12) To verify the identity $\#\mathcal{M}(q, p) = [e_q^k : e_p^{k-1}]$ directly for the function constructed in Theorem 4.12, first note that if $\lambda_q - \lambda_p = 1$, then there is exactly one gradient flow line from q to p if $p \in \overline{W^u(q)}$ and no gradient flow lines from q to p otherwise.

If there are no gradient flow lines from q to p, then $e_p^{k-1} \not< e_q^k$ and the identity is trivial. If there is one gradient flow line from q to p, then the incidence number $[e_q^k : e_p^{k-1}] = \pm 1$, since the CW-complex is regular and $e_p^{k-1} < e_q^k$. So, the identity reduces to a claim concerning compatible orientations in this case.

The sign $\#\mathcal{M}(q, p)$ is determined by choosing orientations on $W^u(q)$, $W^u(p)$, orienting $W(q, p)$ via the short exact sequence

$$0 \longrightarrow T_*W(q, p) \hookrightarrow T_*W^u(q)|_{W(q,p)} \longrightarrow \nu_*(W(q, p), W^u(q))|_{W(q,p)} \longrightarrow 0$$

where the fibers of the normal bundle are isomorphic to $T_p W^u(p)$ via the gradient flow, and then setting $\#\mathcal{M}(q, p) = \pm 1$ depending on whether or not the resulting orientation on $W(q, p)$ agrees with the orientation given by $-\nabla f$. In other words, the vector field $-\nabla f$ determines an outward pointing normal vector on the boundary of the closed k-disk $\overline{W^u(q)}$ at $p \in \partial \overline{W^u(q)}$, and $\#\mathcal{M}(q, p) = \pm 1$ depending on whether or not this outward pointing normal vector followed by a positive basis for $T_p^u W^u(p)$ gives a positive basis for $T_p \overline{W^u(q)}$. Since $\partial \overline{W^u(q)} \approx S^{k-1}$ is oriented similarly using an outward pointing normal vector (see Sect. 2.2), the incidence number $[e_q^k : e_p^{k-1}]$ is also ± 1 following the same rule, cf. Section 2.3 of [8] for the definition of incidence numbers.

Example 4.17 (A Regular CW-Structure on a Real Projective Space) The following diagram illustrates the gradient flow of the Morse-Smale function con-

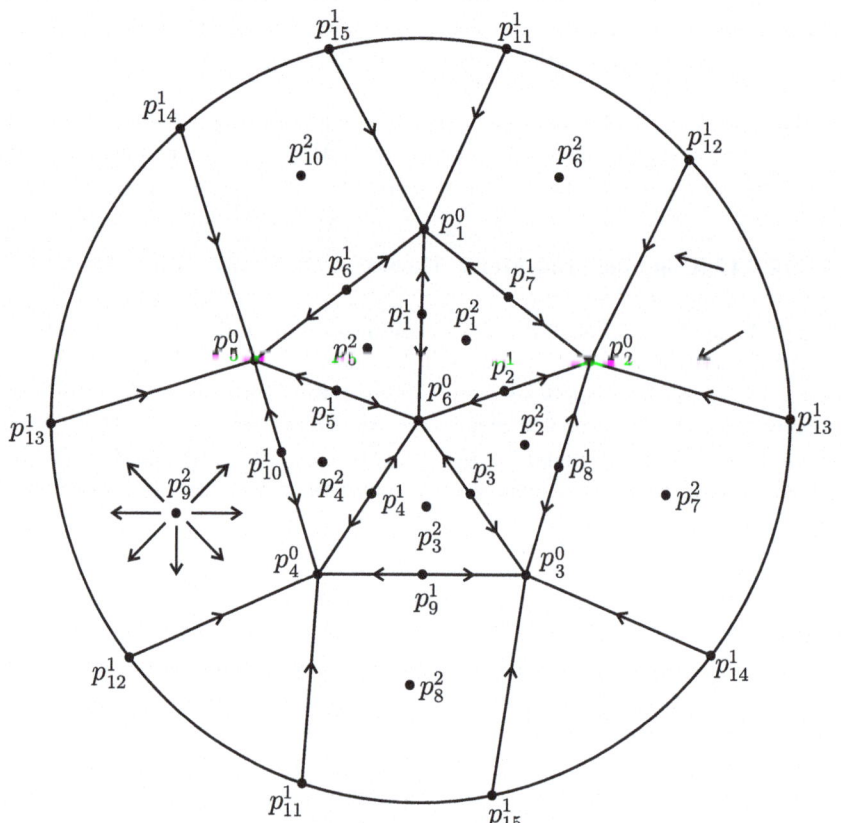

A regular CW-structure determined by unstable manifolds on $\mathbb{R}P^2$

structed in Theorem 4.12 for a minimal triangulation of $\mathbb{R}P^2$ with ten 2-simplices. The triangulation in the diagram is one of two possible irreducible triangulations of $\mathbb{R}P^2$, cf. Figure 3 in [11]. As in Example 2.15, diametrically opposed points on the boundary of the closed disk are identified.

Note that the five 2-simplices in the center of the diagram can be combined to create a single 2-cell. Moreover, the proof of Theorem 4.12 shows that the resulting regular CW-structure with six 2-cells is determined by the unstable manifolds of a Morse-Smale function on $\mathbb{R}P^2$.

Proof of Theorem 4.1 By the Invariance Theorem (Theorem 3.9) and Theorem 4.12 we may assume that (f, g) is a Morse-Smale pair whose unstable manifolds determine a regular CW-structure X on M. Thus, by Lemmas 4.8 and 4.10 we have

$$H_k(M; G) \approx H_k((CW_*(X; G), \partial_*)) \approx H_k((C_*(f; G), \partial_*^G))$$

for all $k = 0, \ldots, m$. □

4.5 Local Coefficient Systems of R-Modules and the Euler Number

The Twisted Morse Homology Theorem (Theorem 4.1) was proved under the assumption that G is a bundle of abelian groups over M, cf. Definition 2.1. However, some local coefficient systems have additional algebraic structures that may be of interest, cf. Remark 2.7. For instance, the fiber of the bundle might be a ring, a module over a ring, a vector space, or a field. In each case, the isomorphisms defining the local system should be isomorphisms in the appropriate category.

In this section we will consider bundles $\mathcal{L}$ of free R-modules over M for some commutative ring R with unit. That is, let M be a topological space, and assume that for every point $x \in M$ we have a free R-module $\mathcal{L}_x$, and for every path $\gamma : [0, 1] \to M$ we have an R-module isomorphism $\gamma_* : \mathcal{L}_{\gamma(1)} \to \mathcal{L}_{\gamma(0)}$ that satisfies the conditions listed in Definition 2.1. This additional algebraic structure allows us to consider the **rank** of the fiber $\mathcal{L}_x$ as a module over R, cf. Theorem 5.1 of [71].

In general, a submodule of a free module might not be free. So, we will assume that R is a principal ideal domain. This assumption guarantees that every submodule of a free R-module C is itself free, and the rank of the submodule is less than or equal to the rank of C, cf. Theorem 6.5 of [71]. In general, the rank of a finitely generated R-module C is defined to be the rank of C/C_{tor}, i.e. the rank of the module modulo its torsion submodule, cf. Theorem 6.8 of [71].

Lemma 4.18 *Let R be a principal ideal domain, assume that C and D are finitely generated free R-modules, and let $\partial : C \to D$ be an R-module homomorphism. Then,*

$$\operatorname{rank} C = \operatorname{rank} \ker \partial + \operatorname{rank} \operatorname{im} \partial.$$

Moreover, if $C' \subseteq C$, then

$$\operatorname{rank} C = \operatorname{rank} C' + \operatorname{rank} C/C'.$$

Proof For any commutative ring with unit, Theorem 5.6 of [71] implies that

$$C \approx \ker \partial \oplus \operatorname{im} \partial.$$

Since R is a principal ideal domain and C is free and finitely generated, $\ker \partial$ is free and finitely generated, cf. Theorem 6.5 of [71]. Similarly, $\operatorname{im} \partial$ is free and finitely generated. Therefore,

$$\operatorname{rank} C = \operatorname{rank} \ker \partial + \operatorname{rank} \operatorname{im} \partial.$$

Now consider the quotient map $C \to C/C'$. Since C is finitely generated, the quotient module C/C' is also finitely generated. Therefore,

$$\frac{C/C'}{(C/C')_{\text{tor}}}$$

is finitely generated and free since R is a principal ideal domain, cf. Theorem 6.8 of [71]. Moreover,

$$\operatorname{rank} \ker \left(C \to C/C' \right) = \operatorname{rank} \ker \left(C \to \frac{C/C'}{(C/C')_{\text{tor}}} \right),$$

because for any basis $\{c_1, \ldots, c_j\}$ for the kernel of the homomorphism on the right there are elements $r_1, \ldots, r_j \in R$ such that $\{r_1 c_1, \ldots, r_j c_j\}$ is a basis for the kernel of the homomorphism on the left. Thus, applying the first assertion of the lemma to the R-module homomorphism

$$C \to \frac{C/C'}{(C/C')_{\text{tor}}}$$

yields the second assertion. □

4.5 Local Coefficient Systems of R-Modules and the Euler Number

Theorem 4.19 (Euler-Poincaré Theorem) *Let R be a principal ideal domain, and let*

$$0 \xrightarrow{\partial_{m+1}} C_m \xrightarrow{\partial_m} C_{m-1} \xrightarrow{\partial_{m-1}} \cdots \xrightarrow{\partial_2} C_1 \xrightarrow{\partial_1} C_0 \xrightarrow{\partial_0} 0$$

be a finite chain complex of finitely generated free R-modules, i.e. assume that C_k is a finitely generated free R-module and ∂_k is an R-module homomorphism for all k. Then

$$\sum_{k=0}^{m}(-1)^k \operatorname{rank} C_k = \sum_{k=0}^{m}(-1)^k \operatorname{rank} H_k((C_*, \partial_*)).$$

Proof By Lemma 4.18 we have

$$\operatorname{rank} C_k = \operatorname{rank} \ker \partial_k + \operatorname{rank} \operatorname{im} \partial_k$$

and

$$\operatorname{rank} H_k((C_*, \partial_*)) = \operatorname{rank} \ker \partial_k - \operatorname{rank} \operatorname{im} \partial_{k+1}$$

for all k. Hence, the two sums are equal since im ∂_{m+1} and im ∂_0 are both 0. □

Definition 4.20 Let X be a finite regular CW-complex of dimension m, and let $\mathcal{L}$ be a bundle of finitely generated free R-modules over X, where R is a principal ideal domain. The $\mathcal{L}$**-twisted Euler number** $\mathcal{X}_\mathcal{L}(X)$ is defined to be

$$\mathcal{X}_\mathcal{L}(X) = \sum_{k=0}^{m}(-1)^k \operatorname{rank} H_k(X; \mathcal{L}),$$

where rank $H_k(X; \mathcal{L})$ denotes the rank of the kth singular homology group of X with coefficients in the bundle $\mathcal{L}$ as a module over R.

Theorem 4.21 (Invariance of the Twisted Euler Number) *If X is a finite regular CW-complex and $\mathcal{L}$ is a bundle of finitely generated free R-modules over X, where R is a principal ideal domain, then the $\mathcal{L}$-twisted Euler number is well-defined, and*

$$\mathcal{X}_\mathcal{L}(X) = \sum_{k=0}^{m}(-1)^k \operatorname{rank} H_k(X; \mathcal{L}) = \sum_{k=0}^{m}(-1)^k \operatorname{rank} H_k(X; \mathcal{L}_{x_0}) = \mathcal{X}_{\mathcal{L}_{x_0}}(X),$$

where $x_0 \in X$ is any basepoint and m is the dimension of X. That is, the $\mathcal{L}$-twisted Euler number is the same as the (untwisted) Euler number for homology with coefficients in the fiber $\mathcal{L}_{x_0}$.

Proof Since X is a finite CW-complex and $\mathcal{L}$ is a bundle of finitely generated free R-modules, the CW-chain group

$$CW_k(X; \mathcal{L}) = \bigoplus_{e_\sigma^k} \mathcal{L}_{x(e_\sigma^k)}$$

is a finitely generated free R-module for all k, where the sum runs over the k-cells e_σ^k in X. By Lemma 4.8, the singular boundary operator with coefficients in $\mathcal{L}$ induces Steenrod's cellular boundary operator with coefficients in $\mathcal{L}$ on a regular CW-complex. Moreover, it's clear that Steenrod's cellular boundary operator is an R-module homomorphism when the isomorphisms that $\mathcal{L}$ associates to paths are R-module homomorphisms, cf. Definition 4.6. Similarly, the singular boundary operator with coefficients in $\mathcal{L}$ is an R-module homomorphism, cf. Definition 4.2. Therefore, the R-module homology of Steenrod's CW-chain complex with coefficients in $\mathcal{L}$ is isomorphic to the R-module singular homology of M with coefficients in $\mathcal{L}$, i.e.

$$H_k((CW_*(X; \mathcal{L}), \partial_*)) \approx H_k(X; \mathcal{L})$$

for all $k = 0, \ldots, m$. This shows that $H_k(X; \mathcal{L})$ is a finitely generated R-module for all k, and therefore $\mathcal{X}_\mathcal{L}(X)$ is well-defined.

The invariance of the $\mathcal{L}$-twisted Euler number follows from the Euler-Poincaré Theorem (Theorem 4.19). That is,

$$CW_k(X; \mathcal{L}) = \bigoplus_{e_\sigma^k} \mathcal{L}_{x(e_\sigma^k)} \approx \bigoplus_{e_\sigma^k} \mathcal{L}_{x_0} = CW_k(X; \mathcal{L}_{x_0})$$

for all $k = 0, \ldots, m$, and therefore

$$\mathcal{X}_\mathcal{L}(X) = \sum_{k=0}^{m}(-1)^k \text{rank } CW_k(X; \mathcal{L}) = \sum_{k=0}^{m}(-1)^k \text{rank } CW_k(X; \mathcal{L}_{x_0}) = \mathcal{X}_{\mathcal{L}_{x_0}}(X).$$

□

Corollary 4.22 *If X is a finite regular CW-complex and $\mathcal{L}$ is a bundle of finitely generated free R-modules of rank one over X, where R is a principal ideal domain, then*

$$\mathcal{X}_\mathcal{L}(X) = \mathcal{X}(X),$$

the classical Euler number of X.

4.5 Local Coefficient Systems of R-Modules and the Euler Number

Proof By the Invariance of the Twisted Euler Number (Theorem 4.21), the Euler-Poincaré Theorem (Theorem 4.19), and the assumption that the fiber $\mathcal{L}_{x_0}$ has rank one we have

$$\mathcal{X}_\mathcal{L}(X) = \mathcal{X}_{\mathcal{L}_{x_0}}(X) = \sum_{k=0}^m (-1)^k \text{rank } CW_k(X; \mathcal{L}_{x_0}) = \sum_{k=0}^m (-1)^k |X^{(k)}|,$$

where $|X^{(k)}|$ denotes the number of k-cells in X. The rightmost sum is $\mathcal{X}(X)$. □

Every closed finite dimensional smooth manifold has a regular CW-structure, cf. Theorem 4.12. Thus, the previous corollary and the classical Morse inequalities, cf. Theorem 3.33 of [8], yield the following.

Corollary 4.23 *Let $f : M \to \mathbb{R}$ be a smooth Morse-Smale function on a closed finite dimensional smooth Riemannian manifold (M, g), and let $\mathcal{L}$ be a bundle of rank one R-modules over M, where R is a principal ideal domain. Then*

$$\mathcal{X}_\mathcal{L}(M) = \sum_{k=0}^m (-1)^k v_k,$$

where v_k denotes the number of critical points of f of index k.

Remark 4.24 The rest of the Morse inequalities, cf. Theorem 3.33 of [8], also hold for the twisted Betti numbers

$$b_k(\mathcal{L}) \stackrel{\text{def}}{=} \text{rank } H_k(M; \mathcal{L})$$

when $\mathcal{L}$ is a bundle of finitely generated free R-modules of rank one on a closed finite dimensional smooth manifold M. This follows from The Twisted Morse Homology Theorem (Theorem 4.1), The Euler-Poincaré Theorem (Theorem 4.19), and the proof of Theorem 3.33 in [8].

Chapter 5
Twisted Morse Cohomology and Lichnerowicz Cohomology

The focus of this chapter is on proving that the η-twisted Morse cohomology groups are isomorphic to the Lichnerowicz cohomology groups associated to $-\eta$. We define the twisted Morse-Smale-Witten cochain complex for a general bundle of abelian groups G, and then we restrict to the case where G is defined by a closed 1-form. In this case, the twisted Morse-Smale-Witten coboundary operator is the usual Morse-Smale-Witten coboundary operator with extra coefficients given by integrating the closed 1-form along the gradient flow lines. We discuss locally conformal symplectic manifolds and Lichnerowicz cohomology, and we prove twisted Morse theoretic versions of the Poincaré Lemma and the de Rham Theorem. Relationships with sheaf cohomology are discussed.

5.1 Twisted Morse Cohomology

Each of the chain complexes defined in the previous chapters has an associated cochain complex, e.g. singular cochain complexes with coefficients in a bundle of abelian groups G [81], Steenrod's CW-cochain complexes with coefficients in G (for regular CW-complexes) [84, Section 31], and Morse-Smale-Witten cochain complexes with coefficients in G.

Definition 5.1 (Twisted Morse-Smale-Witten Cochain Complex) Let $f : M \to \mathbb{R}$ be a smooth Morse-Smale function on a closed smooth Riemannian manifold (M, g) of dimension $m < \infty$. Fix orientations on the unstable manifolds of (f, g), and let G be a bundle of abelian groups over M. For any $k = 0, \ldots, m$, a **Morse-Smale-Witten k-cochain with coefficients in G** is defined to be a function θ that assigns to each critical point $p \in Cr_k(f)$ an element $\theta(p) \in G_p$. The kth **Morse-Smale-Witten cochain group** is the collection of k-cochains, where the group operation is pointwise application of the group operation in G_p. Hence,

$$C^k(f;G) \approx \bigoplus_{p \in Cr_k(f)} G_p.$$

The **Morse-Smale-Witten cochain complex with coefficients in** G is the chain complex $(C^*(f;G), \delta_*^G)$ where $\delta_k^G : C^k(f;G) \to C^{k+1}(f;G)$ is defined on a k-cochain $\theta \in C^k(f;G)$ by

$$(\delta_k^G \theta)(q) = \sum_{p \in Cr_k(f)} \sum_{v \in M(q,p)} \epsilon(v)(\gamma_v)_*(\theta(p)) \in G_q,$$

for any critical point $q \in Cr_{k+1}(f)$, where $\gamma_v : [0,1] \to M$ is any continuous path from q to p whose image coincides with the image of $v \in M(q,p)$ and $\epsilon(v) = \pm 1$ is the sign determined by the orientation on $M(q,p)$.

Remark 5.2 As in Definition 2.11, $-1 \cdot g$ denotes the inverse of $g \in G_q$. The proof that $(\delta_*^G)^2 = 0$ is similar to the proof of Lemma 2.13.

Remark 5.3 The proof of Theorem 3.9 (Invariance Theorem) can be adapted to show that the homology of the Morse-Smale-Witten cochain complex with coefficients in G is independent of the Morse-Smale pair (f, g) and depends only on the isomorphism class of G. We leave the details of showing that the homology of $(C^*(f;G), \delta_*^G)$ does not depend on the Morse-Smale pair (f, g) to the reader. However, the following straightforward proposition, whose proof can be adapted for the twisted Morse-Smale-Witten chain complex, gives an independent proof of the fact that the homology of $(C^*(f;G), \delta_*^G)$ is the same for all bundles of abelian groups in the same isomorphism class.

Proposition 5.4 *If G_1 and G_2 are isomorphic bundles of abelian groups over M, then $H_k((C^*(f;G_1), \delta_*^{G_1})) \approx H_k((C^*(f;G_2), \delta_*^{G_2}))$ for all $k = 0, \ldots, m$.*

Proof By Definition 2.2 there is a family of isomorphisms $\Phi : G_1 \to G_2$ that commute with the homomorphisms associated to a path. Thus, for every $k = 0, \ldots, m$ there is an induced isomorphism $\Phi : C^k(f;G_1) \to C^k(f;G_2)$ given by mapping $\theta \in C^k(f;G_1)$ to $\Phi \circ \theta \in C^k(f;G_2)$ and $\theta \in C^k(f;G_2)$ to $\Phi^{-1} \circ \theta \in C^k(f;G_1)$. Moreover, $\Phi : (C^*(f;G_1), \delta_*^{G_1}) \to (C^*(f;G_2), \delta_*^{G_2})$ is a chain map, because for all $\theta \in C^k(f;G_1)$ and $q \in Cr_{k+1}(f)$ we have

$$(\delta_k^{G_2}(\Phi \circ \theta))(q) = \sum_{p \in Cr_k(f)} \sum_{v \in M(q,p)} \epsilon(v)(\gamma_v)_*((\Phi \circ \theta)(p))$$

$$= \Phi \left(\sum_{p \in Cr_k(f)} \sum_{v \in M(q,p)} \epsilon(v)(\gamma_v)_*(\theta(p)) \right)$$

$$= \Phi((\delta_k^{G_1} \theta)(q)) \in (G_2)_q.$$

5.1 Twisted Morse Cohomology

A similar computation shows that $\Phi^{-1}: (C^*(f; G_2), \delta_*^{G_2}) \to (C^*(f; G_1), \delta_*^{G_1})$ is a chain map, and hence Φ is a chain equivalence. □

Let G be a bundle of abelian groups over a closed connected smooth Riemannian manifold (M, g) of dimension $m < \infty$, let $f : M \to \mathbb{R}$ be a smooth Morse-Smale function on M, and choose a basepoint $x_0 \in Cr_0(f)$ of M. Recall that if M is connected, then the isomorphism class of G is determined by a representation

$$\pi_1(M, x_0) \times G_{x_0} \to G_{x_0},$$

and let $H_{x_0}^* \subset G_{x_0}$ be the subgroup defined by

$$H_{x_0}^* = \{g \in G_{x_0} |\ \gamma_*(g) = g \text{ for all } [\gamma] \in \pi_1(M, x_0)\}.$$

The following theorem gives the 0-dimensional twisted Morse cohomology group of M in terms of the above action, cf. Theorem VI.3.2 of [93].

Theorem 5.5 *If M is connected, then the 0-dimensional twisted Morse cohomology group of M is isomorphic to $H_{x_0}^*$, i.e.*

$$H_0((C^*(f; G), \delta_*^G)) \approx H_{x_0}^*.$$

Proof Let θ be a Morse-Smale-Witten 0-cochain with coefficients in G, i.e. a function that assigns to each $p \in Cr_0(f)$ an element $\theta(p) \in G_p$. If $\delta_0^G \theta = 0$, then for all $q \in Cr_1(f)$ we have

$$(\delta_0^G \theta)(q) = (\gamma_{v_+})_*(\theta(p_+)) - (\gamma_{v_-})_*(\theta(p_-)) = 0 \in G_q,$$

where γ_{v_+} is a path from q to the positive end p_+ of $\overline{W^u(q)}$ and γ_{v_-} is a path from q to the negative end p_- of $\overline{W^u(q)}$. In other words, the value of $\theta(p_+) \in G_{p_+}$ is determined by the value $\theta(p_-) \in G_{p_-}$ via the homomorphism associated to a path in $\overline{W^u(q)}$ from p_+ to p_-.

Since M is connected, every critical point $p \in Cr_0(f)$ can be connected to the basepoint $x_0 \in Cr_0(f)$ by a path contained in the 1-skeleton of f (see the proof of Theorem 2.16). Hence, if $\theta \in \ker \delta_0^G$, then $\theta(p)$ is determined by $\theta(x_0)$ for all $p \in Cr_0(f)$. Moreover, every loop based at x_0 can be homotoped through loops based at x_0 to a loop based at x_0 contained in the 1-skeleton of f, and hence $\gamma_*(\theta(x_0)) = \theta(x_0) \in G_{x_0}$ for all $[\gamma] \in \pi_1(M, x_0)$.

This shows that mapping $\theta \in \ker \delta_0^G$ to $\theta(x_0)$ gives an injective homomorphism $\ker \delta_0^G \to H_0^*$. To see that this map is surjective, simply note that if $g \in G_{x_0}$ satisfies $\gamma_*(g) = g$ for all $[\gamma] \in \pi_1(M, x_0)$, then there is a unique Morse-Smale-Witten 0-cochain $\theta \in \ker \delta_0^G$ such that $\theta(x_0) = g$. □

In this chapter we are mainly interested in cohomology with coefficients in a flat line bundle e^η determined by a closed 1-form η. Note that in this case, e^η is a bundle of rings over M with $e_q^\eta \approx \mathbb{R}$ for all $q \in M$.

Definition 5.6 (η-Twisted Morse-Smale-Witten Cochain Complex) Let $f : M \to \mathbb{R}$ be a smooth Morse-Smale function on a closed finite dimensional smooth Riemannian manifold (M, g). Fix orientations on the unstable manifolds of (f, g), and let $\eta \in \Omega_{cl}^1(M, \mathbb{R})$. The **$\eta$-twisted Morse-Smale-Witten cochain complex** is the chain complex $(C^*(f; e^\eta), \delta_*^\eta)$, where $\delta_k^\eta : C^k(f; e^\eta) \to C^{k+1}(f; e^\eta)$ is defined on a k-cochain $\theta \in C^k(f; e^\eta)$ by

$$(\delta_k^\eta \theta)(q) = \sum_{p \in Cr_k(f)} \sum_{v \in \mathcal{M}(q,p)} \epsilon(v) \exp\left(\int_{\overline{\mathbb{R}}} (\gamma^v)^*(\eta)\right) \theta(p) \in e_q^\eta,$$

for any critical point $q \in Cr_{k+1}(f)$, where $\gamma^v : \overline{\mathbb{R}} \to M$ is any continuous path from p to q whose image coincides with the image of $v \in \mathcal{M}(q, p)$ and $\epsilon(v) = \pm 1$ is the sign determined by the orientation on $\mathcal{M}(q, p)$.

As a consequence of Propositions 2.6 and 5.4 we have the following.

Corollary 5.7 *If $\eta_1, \eta_2 \in \Omega_{cl}^1(M, \mathbb{R})$ are in the same de Rham cohomology class, then $H_k((C^*(f; e^{\eta_1}), \delta_*^{\eta_1})) \approx H_k((C^*(f; e^{\eta_2}), \delta_*^{\eta_2}))$ for all $k = 0, \ldots, m$.*

5.2 Lichnerowicz Cohomology and LCS Manifolds

A closed 1-form $\eta \in \Omega_{cl}^1(M, \mathbb{R})$ on a finite dimensional smooth manifold M can be used to deform the differential of the de Rham cochain complex as follows. For any k-form $\xi \in \Omega^k(M, \mathbb{R})$ define

$$d_\eta \xi = d\xi + \eta \wedge \xi.$$

It is easy to verify that $d_\eta \circ d_\eta = 0$, and hence d_η defines a cochain complex

$$\Omega^0(M, \mathbb{R}) \xrightarrow{d_\eta} \Omega^1(M, \mathbb{R}) \xrightarrow{d_\eta} \Omega^2(M, \mathbb{R}) \xrightarrow{d_\eta} \cdots$$

called the **Lichnerowicz cochain complex**. The homology of this complex is the **Lichnerowicz cohomology**, denoted by $H_\eta^*(M)$; it is sometimes referred to as the **adapted cohomology** of the pair (M, η) [7, 35, 60, 86].

5.2 Lichnerowicz Cohomology and LCS Manifolds

Remark 5.8 The differential d_η in the Lichnerowicz cochain complex can be viewed as a generalization of the Witten deformation [96] to closed 1-forms. That is, if $\eta = dh$ is exact, then $d_\eta \xi = e^{-h} d(e^h \xi)$ for all $\xi \in \Omega^*(M, \mathbb{R})$.

Lichnerowicz cohomology is an invariant used to study locally conformal symplectic manifolds and locally conformal (almost) Kähler manifolds [85, 87].

Definition 5.9 A **locally conformal symplectic (LCS)** form Ω on a finite dimensional smooth manifold M is a smooth nondegenerate 2-form such that there exists an open cover $\mathcal{U} = \{U_i\}_{i \in I}$ of M and smooth positive functions $\lambda_i > 0$ on each U_i such that $\lambda_i \Omega|_{U_i}$ is a symplectic form on U_i, i.e. $\lambda_i \Omega|_{U_i}$ is closed.

A **locally conformal almost Kähler manifold** is a triple (M, J, g), where M is a finite dimensional smooth manifold, J is a smooth almost complex structure on M, and g is a Hermitian Riemannian metric on M, such that the 2-form defined by $\Omega(X, Y) = \mathsf{g}(X, JY)$ is LCS.

Two LCS forms Ω and Ω' on M are said to be **conformally equivalent** if and only if there exists a smooth positive function $h > 0$ such that $\Omega' = h\Omega$.

The connection between the conformal equivalence class of an LCS form Ω and the de Rham cohomology class of an associated closed 1-form $\eta \in \Omega^1_{cl}(M, \mathbb{R})$ is given by the following two propositions [50, 85].

Proposition 5.10 *Let (M, Ω) be an LCS manifold with an open cover $\mathcal{U} = \{U_i\}_{i \in I}$ and associated smooth positive functions $\lambda_i > 0$ such that $\lambda_i \Omega|_{U_i}$ is a symplectic form. Then the forms $\{d(\ln \lambda_i)\}_{i \in I}$ fit together to give a smooth closed 1-form η such that*

$$d\Omega = -\eta \wedge \Omega,$$

and η is uniquely determined by the nondegenerate 2-form Ω.

Conversely, if Ω is a nondegenerate 2-form on a smooth manifold M such that $d\Omega = -\eta \wedge \Omega$ for some closed 1-form η, then Ω is LCS.

Proof Assume $\Omega_i = \lambda_i \Omega|_{U_i}$ is closed. Then $d\lambda_i \wedge \Omega|_{U_i} + \lambda_i d\Omega|_{U_i} = 0$, and hence

$$d\Omega|_{U_i} = -d(\ln \lambda_i) \wedge \Omega|_{U_i}.$$

Moreover, if $U_i \cap U_j \neq \emptyset$ we have

$$d\Omega|_{U_i \cap U_j} = -d(\ln \lambda_i) \wedge \Omega|_{U_i \cap U_j} = -d(\ln \lambda_j) \wedge \Omega|_{U_i \cap U_j},$$

which implies that $d(\ln \lambda_i) = d(\ln \lambda_j)$ since Ω is nondegenerate. Thus, the forms $\{d(\ln \lambda_i)\}_{i \in I}$ fit together to give a closed 1-form η such that $d\Omega = -\eta \wedge \Omega$.

Conversely, if there exists some $\eta \in \Omega^1_{cl}(M, \mathbb{R})$ such that $d\Omega = -\eta \wedge \Omega$, then on any contractible open neighborhood U_i we have $\eta|_{U_i} = df_i$ for some function f_i, and setting $\lambda_i = e^{f_i}$ we have $\eta|_{U_i} = d(\ln \lambda_i)$. Moreover,

$$\begin{aligned}
d(\lambda_i \Omega|_{U_i}) &= d\lambda_i \wedge \Omega|_{U_i} + \lambda_i \, d\Omega|_{U_i} \\
&= d\lambda_i \wedge \Omega|_{U_i} + \lambda_i(-\eta|_{U_i} \wedge \Omega|_{U_i}) \\
&= d\lambda_i \wedge \Omega|_{U_i} - \lambda_i(d(\ln \lambda_i) \wedge \Omega|_{U_i}) \\
&= (d\lambda_i - \lambda_i(d\lambda_i/\lambda_i)) \wedge \Omega|_{U_i} \\
&= 0.
\end{aligned}$$

□

Definition 5.11 The smooth closed 1-form η in the preceding proposition is called the **Lee form** associated to the LCS 2-form Ω.

Proposition 5.12 *If Ω is an LCS form on a finite dimensional smooth manifold M with associated Lee form η and $\Omega' = h\Omega$ for some smooth positive function $h > 0$, i.e. Ω' is conformally equivalent to Ω, then the Lee form associated to Ω' is $\eta - d(\ln h)$. Thus, the de Rham cohomology class of the Lee form $[\eta] \in H^*_{dR}(M; \mathbb{R})$ is an invariant of the conformal class of Ω.*

Proof

$$\begin{aligned}
d\Omega' &= dh \wedge \Omega + h \, d\Omega \\
&= dh \wedge \Omega + h(-\eta \wedge \Omega) \\
&= \frac{dh}{h} \wedge h\Omega - \eta \wedge h\Omega \\
&= -(\eta - d(\ln h)) \wedge \Omega'
\end{aligned}$$

□

Corollary 5.13 *Let (M, Ω) be a closed, smooth, finite dimensional LCS manifold with Lee form $\eta \in \Omega^1_{cl}(M, \mathbb{R})$. Then the η-twisted Morse homology groups $H_*((C_*(f) \otimes \mathbb{R}, \partial^\eta_*))$ and the η-twisted Morse cohomology groups $H_*((C^*(f; e^\eta), \delta^\eta_*))$ are invariants of the conformal class of Ω.*

Proof For homology this follows from Corollary 3.10. For cohomology this follows from Corollary 5.7. □

We now prove the following generalization of Proposition 4.4 of [25], which shows that the Lichnerowicz cohomology groups $H^*_\eta(M)$ are also an invariant of the conformal class of the LCS manifold (M, Ω).

5.2 Lichnerowicz Cohomology and LCS Manifolds

Theorem 5.14 *For any finite dimensional smooth manifold M, the Lichnerowicz cohomology groups $H_\eta^*(M)$ depend only on the cohomology class $[\eta] \in H_{dR}^*(M; \mathbb{R})$. In particular, if η is exact then the Lichnerowicz cohomology groups are isomorphic to the de Rham cohomology groups, i.e. $H_\eta^k(M) \approx H_{dR}^k(M; \mathbb{R})$ for all $k = 0, \ldots, m$.*

Proof First note that every smooth exact 1-form df can be written as dh/h for some smooth positive function $h > 0$ by setting $h = e^f$, because $df = d(\ln h) = dh/h$. So, if η and η' are closed 1-forms on M with $[\eta] = [\eta'] \in H_{dR}^1(M; \mathbb{R})$, then there exists a smooth positive function $h : M \to \mathbb{R}$ such that $\eta' = \eta + dh/h$.

Now, define $\phi : \Omega^*(M, \mathbb{R}) \to \Omega^*(M, \mathbb{R})$ by $\phi(\xi) = \xi/h$ and $\psi : \Omega^*(M, \mathbb{R}) \to \Omega^*(M, \mathbb{R})$ by $\psi(\xi) = h\xi$. Clearly, ϕ and ψ are isomorphisms on $\Omega^k(M, \mathbb{R})$ for all $k = 0, \ldots, m$ with $\phi \circ \psi = id = \psi \circ \phi$. Moreover, $\phi : (\Omega^*(M, \mathbb{R}), d_\eta) \to (\Omega^*(M, \mathbb{R}), d_{\eta'})$ is a chain map, because for all $\xi \in \Omega^*(M, \mathbb{R})$ we have

$$d_{\eta'}(\phi(\xi)) = d\left(\frac{1}{h}\xi\right) + \left(\eta + \frac{dh}{h}\right) \wedge \frac{1}{h}\xi$$

$$= -\frac{1}{h^2}dh \wedge \xi + \frac{1}{h}d\xi + \eta \wedge \frac{1}{h}\xi + \frac{1}{h^2}dh \wedge \xi$$

$$= \frac{1}{h}(d\xi + \eta \wedge \xi)$$

$$= \phi(d_\eta \xi).$$

Similarly, $\psi : (\Omega^*(M, \mathbb{R}), d_{\eta'}) \to (\Omega^*(M, \mathbb{R}), d_\eta)$ is a chain map, because for all $\xi \in \Omega^*(M, \mathbb{R})$ we have

$$d_\eta(\psi(\xi)) = d(h\xi) + \eta \wedge h\xi$$

$$= dh \wedge \xi + hd\xi + \eta \wedge h\xi$$

$$= h\left(d\xi + \eta \wedge \xi + \frac{dh}{h} \wedge \xi\right)$$

$$= \psi(d_{\eta'}\xi).$$

Thus, ϕ is a chain equivalence with inverse ψ, and

$$\phi_* : H_k(\Omega^*(M, \mathbb{R}), d_\eta) \to H_k(\Omega^*(M, \mathbb{R}), d_{\eta'})$$

is an isomorphism for all $k = 0, \ldots, m$. $\square$

5.3 Mapping Differential Forms to Morse-Smale-Witten Cochains

The focal point of this section is Theorem 5.22 (η-Twisted Morse de Rham Theorem), whose proof requires applying Stokes' Theorem to the closure of the unstable manifolds of a Morse-Smale pair (f, g). There are various forms of Stokes' Theorem which might be applied, e.g. Proposition 5.3 of the paper by Burghelea et al. [18] or Proposition 6 in the Appendix by Laudenbach of [13]. However, those versions of Stokes' Theorem rely on the manifolds with corners structure on the (abstractly) compactified unstable manifolds, e.g. Theorem 4.4 of [18], Proposition 2 in the appendix to [13], or Theorem 3.4 of [66], which are all proved under the assumption that the Riemannian metric is locally trivial with respect to the Morse charts.

Note Qin also proved that for a general Morse-Smale pair (f, g) the unstable manifolds can be compactified as smooth manifolds with corners [67]. However, the standard evaluation map $e : \overline{W^u(p)} \to M$ is not necessarily smooth on the boundary when the Riemannian metric is not locally trivial with respect to the Morse charts (Theorem 7.8 of [67]). So, the above referenced versions of Stokes' Theorem might not hold for a general Morse-Smale pair.

In this section we will circumvent the technical issues surrounding compactifying unstable manifolds as smooth manifolds with corners by using a Morse-Smale pair (f, g) whose unstable manifolds coincide with a smooth triangulation of M (Theorem 4.12). For such a pair, the standard Stokes' Theorem for smooth simplices applies, cf. Theorem 4.7 of [89]. Although this choice of (f, g) may seem very restrictive, the homology of the twisted Morse-Smale-Witten cochain complex $(C^*(f; G), \delta_*^G)$ does not depend on the Morse-Smale pair (f, g) (Remark 5.3). So, the main result in this section (Theorem 5.22) holds for all Morse-Smale pairs.

Remark 5.15 The proof of Theorem 3.9 (Invariance Theorem) and the corresponding result for cohomology rely heavily on the smooth manifolds with corners structure for compactified moduli spaces of gradient flow lines. However, the smooth manifolds with corners structure for moduli spaces of gradient flow lines has been proved for all Morse-Smale pairs (f, g), without the assumption that the metric g is locally trivial with respect to the Morse charts, cf. Theorem 7.5 of [67] or Theorem 1.1 of [91].

Let $f : M \to \mathbb{R}$ be a smooth Morse-Smale function on a closed smooth Riemannian manifold (M, g) of finite dimension m whose unstable manifolds coincide with a smooth triangulation of M (Theorem 4.12). That is, for any critical point $q \in Cr(f)$ the closure of the unstable manifold $U_q \stackrel{\text{def}}{=} \overline{W^u(q)}$ is the

5.3 Mapping Differential Forms to Morse-Smale-Witten Cochains

diffeomorphic image $\sigma : \Delta^{\lambda_q} \to M$ of a λ_q-simplex, where λ_q denotes the index of q. By the λ-lemma we have

$$\partial \overline{W^u(q)} = \bigcup_{q \succ p} W^u(p) \subseteq M,$$

where the notation $q \succ p$ means that there is a non-trivial broken gradient flow line from q to $p \in Cr(f)$, cf. Section 6.3 of [8]. Since the unstable manifolds of (f, g) coincide with the triangulation, $U_p = \overline{W^u(p)}$ is the diffeomorphic image of a face of Δ^{λ_q} under $\sigma : \Delta^{\lambda_q} \to M$ whenever $q \succ p$ and $\lambda_q - \lambda_p = 1$.

Definition 5.16 Fix any $\eta \in \Omega^1_{cl}(M, \mathbb{R})$ and note that for any $q \in Cr(f)$ the restriction $-\eta|_{U_q} = d(\ln h)$ for some smooth positive function $h : U_q \to \mathbb{R}$, since $-\eta|_{U_q}$ is exact and U_q is the diffeomorphic image of a simplex. For any $\xi \in \Omega^k(M, \mathbb{R})$, define $\theta_\xi(q) = \xi(q)$ if $k = 0$ or

$$\theta_\xi(q) = \frac{1}{h(q)} \int_{U_q} h\xi \in e_q^\eta$$

if $k = 1, \ldots, m$. Note that this definition is independent of the choice of h because $-\eta|_{U_q} = d(\ln \tilde{h}) = d(\ln h)$ implies $\tilde{h} = Ch$ for some $C \in \mathbb{R}$. Thus, for any $k = 0, \ldots, m$ a k-form ξ determines a well-defined Morse-Smale-Witten k-cochain θ_ξ, and $F(\xi) = \theta_\xi$ defines a linear map $F : \Omega^k(M, \mathbb{R}) \to C^k(f; e^\eta)$.

Remark 5.17 The map $F : \Omega^k(M, \mathbb{R}) \to C^k(f; e^\eta)$ defined above is similar to the one used to prove the de Rham Theorem, cf. Section V.5 of [17], as well as the map defined in Section 3.4 of [5].

Proposition 5.18 $F : (\Omega^*(M, \mathbb{R}), d_{-\eta}) \to (C^*(f; e^\eta), \delta_*^\eta)$ *is a chain map. That is, the map preserves degree and* $F \circ d_{-\eta} = \delta_*^\eta \circ F$.

Proof Pick any $q \in Cr_{k+1}(f)$, let $-\eta|_{U_q} = d(\ln h)$ for some smooth positive function h on $U_q = \overline{W^u(q)} \approx \Delta^{k+1}$, and note that for any $\xi \in \Omega^k(M, \mathbb{R})$ we have

$$d_{-\eta}\xi = d\xi + \frac{dh}{h} \wedge \xi = \frac{1}{h} d(h\xi)$$

on U_q. Moreover, for any choice of orientations on the unstable manifolds the signs $\epsilon(v) = \pm 1$ satisfy the relation

$$\partial \overline{W^u(q)} = \bigcup_{p \in Cr_k(f)} \bigcup_{v \in \mathcal{M}(q,p)} \epsilon(v) \overline{W^u(p)}$$

as oriented manifolds, cf. Remark 4.16. Hence, $(F \circ d_{-\eta}(\xi))(q) =$

$$\theta_{d_{-\eta}\xi}(q) = \frac{1}{h(q)} \int_{U_q} h d_{-\eta}\xi$$

$$= \frac{1}{h(q)} \int_{U_q} d(h\xi)$$

$$= \frac{1}{h(q)} \int_{\partial U_q} h\xi$$

$$= \frac{1}{h(q)} \sum_{p \in Cr_k(f)} \sum_{v \in \mathcal{M}(q,p)} \epsilon(v) \int_{U_p} h\xi$$

$$= \frac{1}{h(q)} \sum_{p \in Cr_k(f)} \sum_{v \in \mathcal{M}(q,p)} \epsilon(v) h(p) \theta_\xi(p)$$

$$= \sum_{p \in Cr_k(f)} \sum_{v \in \mathcal{M}(q,p)} \epsilon(v) e^{\ln h(p) - \ln h(q)} \theta_\xi(p)$$

$$= \sum_{p \in Cr_k(f)} \sum_{v \in \mathcal{M}(q,p)} \epsilon(v) \exp\left(\int_\mathbb{R} (\gamma^v)^*(-d(\ln h))\right) \theta_\xi(p)$$

$$= \sum_{p \in Cr_k(f)} \sum_{v \in \mathcal{M}(q,p)} \epsilon(v) \exp\left(\int_\mathbb{R} (\gamma^v)^*(\eta)\right) \theta_\xi(p)$$

$$= (\delta_k^\eta \circ F(\xi))(q),$$

were γ^v is any parameterization of v from p to q. □

To prove that the chain map F induces an isomorphism between the Lichnerowicz groups $H^*_{-\eta}(M)$ and the η-twisted Morse cohomology groups of (M, f, g, η), we will use the following local version of the η-twisted Morse-Smale-Witten cochain complex.

Definition 5.19 (Local η-Twisted Morse-Smale-Witten Cochain Complex) For any open set $U \subseteq M$ and for all $k = 0, \ldots, m$ define $C_U^k(f; e^\eta) \subseteq C^k(f; e^\eta)$ to be the k-cochains θ such that $\theta(p) = 0$ for all $p \in Cr_k(f) - U$, and for any $\theta \in C^k(f; e^\eta)$ define $\theta|_U \in C_U^k(f; e^\eta)$ to be

$$\theta|_U(p) = \begin{cases} \theta(p) & \text{if } p \in Cr_k(f) \cap U \\ 0 & \text{if } p \in Cr_k(f) - U. \end{cases}$$

Define $\delta_k^\eta|_U : C_U^k(f; e^\eta) \to C_U^{k+1}(f; e^\eta)$ by restricting the sum defining δ_k^η in Definition 5.6 to the critical points $p \in Cr_k(f) \cap U$ when evaluating $\delta_k^\eta|_U \theta$ on a critical point $q \in Cr_{k+1}(f) \cap U$.

5.3 Mapping Differential Forms to Morse-Smale-Witten Cochains

We will call an open subset $U \subseteq M$ **unstably closed** if for every critical point $p \in Cr(f) \cap U$ we have $\overline{W^u(p)} \subseteq U$. We will call a set $U \subseteq M$ **unstably 0-connected** if for every $p_1, p_2 \in Cr_0(f) \cap U$ there exists $q_1, q_2, \ldots, q_n \in Cr_1(f) \cap U$ such that $\bigcup_{i=1}^{n} \overline{W^u(q_i)}$ is the image of a continuous path from p_1 to p_2.

Remark 5.20 If $U \subseteq M$ is unstably closed, then for all $p, q \in Cr(f) \cap U$ such that $\lambda_q - \lambda_p$ is 1 or 2 we have $Im(\overline{\mathcal{M}(q,p)}) \subset U$, cf. Corollary 6.27 of [8]. Thus, $(\delta_*^\eta|_U)^2 = 0$ whenever U is unstably closed, cf. Lemmas 2.9 and 2.13, and $(C_U^*(f;e^\eta), \delta_*^\eta|_U)$ is a cochain complex. The additional assumption that U is unstably 0-connected implies that $H_0((C_U^*(f;e^\eta), \delta_*^\eta|_U)) \approx \mathbb{R}$.

Lemma 5.21 (η-Twisted Morse Poincaré Lemma) *Let $f : M \to \mathbb{R}$ be a smooth Morse-Smale function on a closed finite dimensional smooth Riemannian manifold $(M, \mathbf{g})$ such that the unstable manifolds of $(f, \mathbf{g})$ coincide with a smooth triangulation of M. If $U \subseteq M$ is an open contractible set that is unstably closed and unstably 0-connected, then*

$$F_* : H^k_{-\eta}(U) \to H_k((C_U^*(f;e^\eta), \delta_*^\eta|_U))$$

is an isomorphism for all $k = 0, \ldots, m$. Moreover, $H^0_{-\eta}(U) \approx \mathbb{R}$ and $H^k_{-\eta}(U) = 0$ for $k = 1, \ldots, m$.

Proof Since U is contractible and η is closed, $H^k_{-\eta}(U) \approx H^k_{dR}(U; \mathbb{R})$ for all k by Theorem 5.14, and the second assertion follows from the usual Poincaré Lemma.

Now, for every $k = 0, \ldots, m$ pick an open neighborhood N_k of $Cr_k(f) \cap U$ consisting of a union of contractible neighborhoods N_k^p around each critical point p of index k in U, small enough so that no component of N_k contains more than one critical point and $N_j \cap N_k = \emptyset$ if $j \neq k$. For all $k = 1, \ldots, m$ let ξ_{k-1} be a smooth $(k-1)$-form with support in N_k such that

$$\frac{1}{h(p)} \int_{U_p} h d_{-\eta} \xi_{k-1} = 1 \in e_p^\eta$$

for all $p \in Cr_k(f) \cap U$, where $U_p = \overline{W^u(p)}$ and $-\eta|_U = d(\ln h)$.

For all $k = 0, \ldots, m$ and every $\theta_k \in C^k(f;e^\eta)$ let $\overline{\theta_k}$ be the extension of θ_k to N_k that is constant on each component N_k^p of N_k, i.e. $\overline{\theta_k}|_{N_k^p} \equiv \theta_k(p)$, and define a linear map $G : C^k(f;e^\eta) \to \Omega^k(M, \mathbb{R})$ by

$$G(\theta_k) = \overline{\theta_k} d_{-\eta} \xi_{k-1} + \overline{\delta_k^\eta \theta_k} \xi_k,$$

where we define $\xi_{-1} = \xi_m = 0$. Note that $G(\theta_k)$ is a smooth k-form, even though $\overline{\theta_k}$ and $\overline{\delta_k^\eta \theta_k}$ are discontinuous, because $d_{-\eta}\xi_{k-1}$ is smooth with support in N_k and ξ_k is smooth with support in N_{k+1}. In other words, for every $k = 1, \ldots, m$, N_k is a disjoint union of contractible open sets $N_k = \bigcup_{p \in Cr_k(f)} N_k^p$, and we can define ξ_{k-1}^p to be the smooth $(k-1)$-form that is equal to ξ_{k-1} on the component N_k^p and zero outside of N_k^p. Then

$$G(\theta_k) = \sum_{p \in Cr_k(f)} \theta_k(p) d_{-\eta} \xi_{k-1}^p + \sum_{q \in Cr_{k+1}(f)} (\delta_k^\eta \theta_k)(q) \xi_k^q,$$

which is a smooth k-form on M.

For all $k = 0, \ldots, m$ and any $\theta_k \in C^k(f; e^\eta)$ we have

$$d_{-\eta}(G(\theta_k)) = d_{-\eta}\left(\overline{\theta_k} d_{-\eta} \xi_{k-1} + \overline{\delta_k^\eta \theta_k} \xi_k\right)$$
$$= \overline{\theta_k} d_{-\eta}^2 \xi_{k-1} + \overline{\delta_k^\eta \theta_k} d_{-\eta} \xi_k$$
$$= \overline{\delta_k^\eta \theta_k} d_{-\eta} \xi_k + \overline{\delta_{k+1}^\eta \delta_k^\eta \theta_k} \xi_{k+1}$$
$$= G(\delta_k^\eta \theta_k).$$

Thus, $d_{-\eta} \circ G = G \circ \delta_*^\eta$, i.e. $G : (C^*(f; e^\eta), \delta_*^\eta) \to (\Omega^*(M, \mathbb{R}), d_{-\eta})$ is a chain map. Moreover, for all $k = 1, \ldots, m$ we have for any $\theta_k \in C^k(f; e^\eta)$

$$((F \circ G)(\theta_k))(p) = F(\overline{\theta_k} d_{-\eta} \xi_{k-1} + \overline{\delta_k^\eta \theta_k} \xi_k)(p)$$
$$= \frac{1}{h(p)} \int_{U_p} h \overline{\theta_k} d_{-\eta} \xi_{k-1}$$
$$= \theta_k(p) \frac{1}{h(p)} \int_{U_p} h d_{-\eta} \xi_{k-1}$$
$$= \theta_k(p) \in e_p^\eta,$$

for all $p \in Cr_k(f)$, and hence

$$(F \circ G)_* : H_k((C_U^*(f; e^\eta), \delta_*^\eta|_U)) \to H_k((C_U^*(f; e^\eta), \delta_*^\eta|_U))$$

is the identity for all $k = 1, \ldots, m$. This implies that

$$F_* : H_{-\eta}^k(M) \to H_k((C_U^*(f; e^\eta), \delta_*^\eta|_U))$$

is surjective for $k = 1, \ldots, m$, and hence F_* is the trivial isomorphism for all $k = 1, \ldots, m$, since $H_{-\eta}^k(M) = 0$.

Now, if $\xi \in \Omega^0(U)$, then

$$d_{-\eta} \xi = d\xi + \frac{dh}{h} \wedge \xi = \frac{1}{h} d(h\xi)$$

implies that $\xi \in \ker d_{-\eta}$ if and only if $\xi = C/h$ for some $C \in \mathbb{R}$, where $-\eta|_U = d(\ln h)$. Moreover, Proposition 5.18 and the assumption that U is unstably closed

5.3 Mapping Differential Forms to Morse-Smale-Witten Cochains

implies that if $\xi \in \ker d_{-\eta}$, then $\theta_\xi = F(\xi)$ satisfies $(\delta_0^\eta \theta_\xi)(q) = 0$ for all $q \in Cr_1(f) \cap U$, i.e. $F : \ker d_{-\eta} \to \ker \delta_0^\eta$. Note that $F(C/h) = C/h|$, where $h|$ denotes the restriction of $h : U \to \mathbb{R}$ to $Cr_0(f) \cap U$.

Define a map $G : \ker \delta_0^\eta \to \ker d_{-\eta}$ by $G(\theta_0) = \theta_0(p)h(p)/h$ for any $p \in Cr_0(f) \cap U$. To see that this definition is independent of $p \in Cr_0(f) \cap U$, first note that if $p_+, p_- \in Cr_0(f) \cap U$ are the boundary points of an unstable manifold $W^u(q)$ with $q \in Cr_1(f) \cap U$, then $\mathcal{M}(q, p_+)$ consists of a single element v_{p_+} and $\mathcal{M}(q, p_-)$ consists of a single element v_{p_-}. Moreover, $\epsilon(v_{p_+}) = -\epsilon(v_{p_-})$, and we can label the points so that $\epsilon(v_{p_+}) = +1$. Thus,

$$0 = (\delta_0^\eta \theta_0)(q)$$
$$= \exp\left(\int_\mathbb{R} (\gamma^{v_{p_+}})^*(\eta)\right) \theta_0(p_+) - \exp\left(\int_\mathbb{R} (\gamma^{v_{p_-}})^*(\eta)\right) \theta_0(p_-)$$
$$= e^{\ln h(p_+) - \ln h(q)} \theta_0(p_+) - e^{\ln h(p_-) - \ln h(q)} \theta_0(p_-)$$
$$= \frac{h(p_+)}{h(q)} \theta_0(p_+) - \frac{h(p_-)}{h(q)} \theta_0(p_-),$$

where $\gamma^{v_{p_+}}$ and $\gamma^{v_{p_-}}$ are paths that end at q. Therefore, $h(p_+)\theta_0(p_+) = h(p_-)\theta_0(p_-)$. The fact that $G(\theta_0) = \theta_0(p)h(p)/h$ does not depend on the choice of $p \in Cr_0(f) \cap U$ now follows inductively using the assumption that U is unstably 0-connected.

So, for any $\theta_0 \in \ker \delta_0^\eta$ we have $(F \circ G)(\theta_0) = F(\theta_0(p)h(p)/h) = \theta_0(p)h(p)/h| = \theta_0$, since $\theta_0(p)h(p)$ does not depend on $p \in Cr_0(f) \cap U$. Moreover, for any $C/h \in \ker d_{-\eta}$ we have $(G \circ F)(C/h) = G(C/h|) = \frac{C}{h(p)} h(p)/h = C/h$, and hence $F \circ G = G \circ F = id$. Therefore, $G : \ker \delta_0^\eta \to \ker d_{-\eta}$ is a well-defined group homomorphism with inverse F, and $F_* : H^0_{-\eta}(M) \to H_0((C_U^*(f; e^\eta), \delta_*^\eta|_U))$ is an isomorphism.

This completes the proof of Lemma 5.21 □

Theorem 5.22 (η-**Twisted Morse de Rham Theorem**) *Let $f : M \to \mathbb{R}$ be a smooth Morse-Smale function on a closed finite dimensional smooth Riemannian manifold (M, g). For any $\eta \in \Omega^1_{cl}(M, \mathbb{R})$, the η-twisted Morse cohomology groups are isomorphic to the Lichnerowicz cohomology groups defined by $-\eta$, i.e.*

$$H_k((C^*(f; e^\eta), \delta_*^\eta)) \approx H^k_{-\eta}(M)$$

for all $k = 0, \ldots, m$.

Proof Pick a Morse-Smale pair (f, g) whose unstable manifolds coincide with a smooth triangulation of M, and fix orientations on the unstable manifolds. Recall that for any two open sets U, V in M there is short exact sequence of cochain complexes

$$0 \longrightarrow \Omega^k(U \cup V) \longrightarrow \Omega^k(U) \oplus \Omega^k(V) \longrightarrow \Omega^k(U \cap V) \longrightarrow 0,$$

where the second map is defined by $\xi \mapsto (\xi|_U, -\xi|_V)$ and the third map is given by $(\xi, \xi') \mapsto \xi|_{U\cap V} + \xi'|_{U\cap V}$. Assuming that U and V are unstably closed, we can define a similar short exact sequence of cochain complexes

$$0 \longrightarrow C^k_{U\cup V}(f; e^\eta) \longrightarrow C^k_U(f; e^\eta) \oplus C^k_V(f; e^\eta) \longrightarrow C^k_{U\cap V}(f; e^\eta) \longrightarrow 0,$$

where the second map is defined by $\theta \mapsto (\theta|_U, -\theta|_V)$ and the third map is given by $(\theta, \theta') \mapsto \theta|_{U\cap V} + \theta'|_{U\cap V}$.

Moreover, the chain map $F : (\Omega^*(M, \mathbb{R}), d_{-\eta}) \to (C^*(f; e^\eta), \delta^\eta_*)$ induces chain maps that fit into the commutative diagram

$$\begin{array}{ccccccccc}
0 & \longrightarrow & \Omega^k(U \cup V) & \longrightarrow & \Omega^k(U) \oplus \Omega^k(V) & \longrightarrow & \Omega^k(U \cap V) & \longrightarrow & 0 \\
& & \downarrow F & & \downarrow F \oplus F & & \downarrow F & & \\
0 & \longrightarrow & C^k_{U\cup V}(f; e^\eta) & \longrightarrow & C^k_U(f; e^\eta) \oplus C^k_V(f; e^\eta) & \longrightarrow & C^k_{U\cap V}(f; e^\eta) & \longrightarrow & 0
\end{array}$$

which in turn induces the following ladder commutative diagram with exact rows.

$$\begin{array}{ccccccccc}
\cdots & \longrightarrow & H^k_{-\eta}(U \cup V) & \longrightarrow & H^k_{-\eta}(U) \oplus H^k_{-\eta}(V) & \longrightarrow & H^k_{-\eta}(U \cap V) & \longrightarrow & \cdots \\
& & \downarrow F_* & & \downarrow F_* \oplus F_* & & \downarrow F_* & & \\
\cdots & \longrightarrow & H^k_{U\cup V}(f; e^\eta) & \longrightarrow & H^k_U(f; e^\eta) \oplus H^k_V(f; e^\eta) & \longrightarrow & H^k_{U\cap V}(f; e^\eta) & \longrightarrow & \cdots
\end{array}$$

The 5-lemma applied to this diagram implies that if F_* is an isomorphism on U, V, and, $U \cap V$ for $k = i - 1$ and i, then F_* is an isomorphism on $U \cup V$ when $k = i$.

Now, note that both the union and intersection of unstably closed sets is unstably closed, and F_* is an isomorphism for all k on every open unstably closed set with components that are contractible and unstably 0-connected by Lemma 5.21. Thus, if V_1 and V_2 are open and unstably closed, and V_1, V_2, and $V_1 \cap V_2$ have components that are contractible and unstably 0-connected, then the above ladder commutative diagram implies that F_* is an isomorphism for all k on $V_1 \cup V_2$. Continuing by induction, assume that we have a finite collection $\{V_j\}_{j=1}^n$ of open unstably closed sets with components that are contractible and unstably 0-connected such that

- $\left(\bigcup_{j=1}^{n-1} V_j\right) \cap V_n$ has components that are contractible and unstably 0-connected
- F_* is an isomorphism for all k on $\bigcup_{j=1}^{n-1} V_j$.

Then taking $U = \bigcup_{j=1}^{n-1} V_j$ and $V = V_n$, Lemma 5.21 and the above ladder commutative diagram imply that F_* is an isomorphism for all k on $\bigcup_{j=1}^n V_j$. This shows that F_* is an isomorphism for all k on any unstably closed set U that can be written as a finite union $U = V_1 \cup \cdots \cup V_n$ of open unstably closed sets $\{V_j\}_{j=1}^n$

5.3 Mapping Differential Forms to Morse-Smale-Witten Cochains

with components that are contractible and unstably 0-connected such that

(∗) $\left(\bigcup_{j=1}^{i-1} V_j\right) \cap V_i$ has components that are contractible and unstably 0-connected for all $i = 2, \ldots, n$.

To complete the proof, take a finite open cover $\{U_j\}_{j=1}^n$ of contractible sets that are unstably closed and unstably 0-connected consisting of a small thickening of $\overline{W^u(p_j)}$ for each critical point p_j of index $m = \dim M$. If $U_1 \cap U_2 \neq \emptyset$, then this intersection can be written as a finite union of open unstably closed sets $\{V_j\}_{j=1}^n$ with components that are contractible and unstably 0-connected satisfying condition (∗) because the unstable manifolds of (f, g) determine a regular CW-structure on M. Thus, Lemma 5.21 and the above ladder commutative diagram imply that F_* is an isomorphism for all k on $U_1 \cup U_2$.

Now assume that we have shown that F_* is an isomorphism for all k on $U_1 \cup \cdots \cup U_{i-1}$ for some $i = 2, \ldots, n$. The set $\left(\bigcup_{j=1}^{i-1} U_i\right) \cap U_i$ can be written as a finite union of open unstably closed sets $\{V_j\}_{j=1}^n$ with components that are contractible and unstably 0-connected satisfying condition (∗) since the unstable manifolds of (f, g) determine a regular CW-structure. Therefore, Lemma 5.21 and the above ladder commutative diagram imply that F_* is an isomorphism for all k on $U_1 \cup \cdots \cup U_i$ and hence on $U_1 \cup \cdots \cup U_n = M$.

This completes the proof of Theorem 5.22, since the homology of $(C^*(f; e^\eta), \delta_*^\eta)$ does not depend on the Morse-Smale pair (f, g), cf. Remark 5.3. □

Remark 5.23 The above proof is modeled on the proof of the de Rham Theorem found in Section V.9 of [17].

Remark 5.24 A result similar to Theorem 5.22 was proved by Bismut and Zhang for flat vector bundles using analysis and geometric measure theory, cf. Section II.c of [13]. Their result relies on results by Laudenbach contained in the appendix to [13], which use the assumption that the Riemannian metric is locally trivial with respect to the Morse charts.

Corollary 5.25 (Invariance of the η-Twisted Euler Number) *Let M be a closed finite dimensional smooth manifold and $\eta \in \Omega^1_{cl}(M, \mathbb{R})$ a closed 1-form on M. The η-twisted Lichnerowicz Euler number*

$$\mathcal{X}^\eta(M) \stackrel{def}{=} \sum_{k=0}^m (-1)^k \dim H_\eta^k(M)$$

is well-defined and equal to the classical Euler number $\mathcal{X}(M)$.

Proof Note that $(\Omega^*(M, \mathbb{R}), d_{-\eta})$ is a complex of $\mathbb{R}$-modules which is not finitely generated, whereas $(C^*(f; e^\eta), \delta_*^\eta)$ is a complex of finitely generated free $\mathbb{R}$-modules. Moreover, the map $F : \Omega^k(M, \mathbb{R}) \to C^k(f; e^\eta)$ from Definition 5.16 is an $\mathbb{R}$-module homomorphism. Therefore, the η-Twisted Morse de Rham Theorem

(Theorem 5.22) implies that $H^k_{-\eta}(M)$ is a finite dimensional real vector space for all k and all $\eta \in \Omega^1_{cl}(M, \mathbb{R})$. This implies that the η-twisted Lichnerowicz Euler number

$$\mathcal{X}^\eta(M) = \sum_{k=0}^m (-1)^k \dim H^k_\eta(M) = \sum_{k=0}^m (-1)^k \dim H_k((C^*(f; e^{-\eta}), \delta_*^{-\eta}))$$

is well-defined. The fact that $\mathcal{X}^\eta(M) = \mathcal{X}(M)$ follows from the η-Twisted Morse de Rham Theorem (Theorem 5.22) and the Euler-Poincaré Theorem (Theorem 4.19), cf. Theorem 4.21. □

Remark 5.26 (Elliptic Operators and Index Theory) Corollary 5.25 can also be proved for the complex $(\Omega^*(M, \mathbb{R}), d_\eta)$ using the index theory of elliptic operators. That is, in Section 6 of their classic paper [3] Atiyah and Singer note that on any closed finite dimensional smooth manifold M the de Rham complex of complex valued differential forms is elliptic, and its Euler characteristic is the usual Euler number. They then consider the operator $d + d^*$, where d^* denotes the adjoint of d with respect to a Riemannian metric on M. Restricting $d + d^*$ to the even dimensional complex valued differential forms gives an elliptic differential operator of order 1 whose index is the Euler number of the de Rham complex, cf. Section III.4.D of [15] and Proposition 19.1.16 of [41]. Perturbing d to d_η is a lower order perturbation, and hence the index of $d + d^*$ does not change under the perturbation. Therefore,

$$\mathcal{X}(M) = \text{index}(d + d^*) = \text{index}(d_\eta + (d_\eta)^*) = \mathcal{X}^\eta(M).$$

5.4 Relationship to Sheaf Cohomology

Proposition 5.18 shows that the map F given in Definition 5.16 is a chain map from the Lichnerowicz cochain complex $(\Omega^*(M, \mathbb{R}), d_{-\eta})$ to the η-twisted Morse-Smale-Witten cochain complex $(C^*(f; e^\eta), \delta_*^\eta)$, and Theorem 5.22 shows that this chain map induces an isomorphism of the corresponding cohomology groups. A result due to Vaisman shows that the Lichnerowicz cohomology groups $H^*_{-\eta}(M)$ are also isomorphic to the cohomology of M with coefficients in a sheaf [86].

Vaisman defines $\mathcal{F}_\eta(M)$ to be the sheaf of germs of differentiable functions $f : M \to \mathbb{R}$ such that

$$d_{-\eta} f = df - \eta f = 0.$$

Letting $\mathcal{A}^k(M)$ be the sheaf of germs of differentiable k-forms on M for all $k = 0, \ldots, m$, he shows that

5.4 Relationship to Sheaf Cohomology

$$0 \longrightarrow \mathscr{F}_\eta(M) \lhook\joinrel\longrightarrow \mathscr{A}^0(M) \xrightarrow{d_{-\eta}} \mathscr{A}^1(M) \xrightarrow{d_{-\eta}} \cdots$$

is a fine resolution of $\mathscr{F}_\eta(M)$ by proving that $d_{-\eta}$ satisfies a Poincaré Lemma. This proves that $H^k(M; \mathscr{F}_\eta(M))$, the kth-cohomology group of M with coefficients in the sheaf $\mathscr{F}_\eta(M)$, is isomorphic to the Lichnerowicz cohomology group $H^k_{-\eta}(M)$ for all $k = 0, \ldots, m$ (Proposition 3.1 of [86]). Applying Theorem 5.22, we see that the η-twisted Morse cohomology group $H_k((C^*(f; e^\eta), \delta^\eta_*))$ is also isomorphic to $H^k(M; \mathscr{F}_\eta(M))$ for all $k = 0, \ldots, m$.

Now, if X is a paracompact Hausdorff space that is locally path connected and semilocally 1-connected, then there is a bijection between equivalence classes of local coefficient systems on X and equivalence classes of locally constant sheaves on X, cf. Proposition 2.5.1 of [26] or Exercises 6.F of [80]. The above cohomology isomorphisms suggest that the equivalence class of the sheaf $\mathscr{F}_\eta(M)$ should correspond to the equivalence class of the local system e^η under this bijection. To prove this, first note that the fact that $\mathscr{F}_\eta(M)$ is locally constant is implicit in Section 3 of [86]; we reformulate this explicitly in the following lemma.

Lemma 5.27 ([86]) *If M is a finite dimensional smooth manifold and U is an open connected subset of M such that $\eta|_U$ is exact, then*

$$\mathscr{F}_\eta(M)|_U \approx \mathbb{R}|_U.$$

Moreover, if $\eta|_U$ is not exact, then the only section of $\mathscr{F}_\eta(M)|_U$ is the zero section.

Proof Suppose that $f : U \to \mathbb{R}$ is a differentiable function such that

$$d_{-\eta} f = df - \eta|_U f = 0.$$

If $\eta|_U$ is exact, then $-\eta|_U = dh/h$ for some positive smooth function $h : U \to \mathbb{R}$ and

$$d_{-\eta} f = df + \frac{dh}{h} f = \frac{1}{h} d(hf) = 0.$$

Therefore, $f = C/h$ for some $C \in \mathbb{R}$.

If $f(x) \neq 0$ for all $x \in U$, then the first equation shows that $\eta|_U = df/f = d(\ln f)$, i.e. $\eta|_U$ is exact. Thus, if $\eta|_U$ is not exact, then $f(x_0) = 0$ for some $x_0 \in U$, which implies that $f \equiv 0$ on any open neighborhood V of x_0 where $\eta|_V$ is exact. For any other point $x_1 \in U$ we can cover a path from x_0 to x_1 with finitely many open sets $\{V_j\}_{j=1}^n$ such that $\eta|_{V_j}$ is exact for all $j = 1, \ldots, n$ and conclude that $f(x_1) = 0$. Therefore, the only section of $\mathscr{F}_\eta(M)|_U$ is the zero section when $\eta|_U$ is not exact. □

Following Exercise 6.F.1 of [80], we now recall that the local system e^η on M determines a presheaf $\mathscr{F}_{e^\eta}$ on M as follows: for any open set $U \subseteq M$, let $\mathscr{F}_{e^\eta}(U)$ be the set of all functions f assigning to each $x \in U$ an element $f(x) \in \mathbb{R}_x$ with the property that for any continuous path $\gamma : [0, 1] \to U$ we have

$$f(\gamma(1)) = e^{\int_0^1 \gamma^*(\eta)} f(\gamma(0)).$$

According to Exercise 6.F.1 of [80], the presheaf $\mathscr{F}_{e^\eta}$ satisfies the two additional axioms necessary to be a sheaf, cf. Section 6.7.5 of [80]. Moreover, if $\eta|_U$ is exact, then $-\eta|_U = d(\ln h)$ for some positive smooth function $h : U \to \mathbb{R}$ and we have

$$f(\gamma(1)) = \frac{h(0) f(\gamma(0))}{h(1)}.$$

That is, $f = C/h$ for $C = h(0) f(\gamma(0))$ on any connected open set U such that $-\eta|_U = d(\ln h)$, and hence $\mathscr{F}_{e^\eta}(U) \approx \mathbb{R}|_U \approx \mathscr{F}_\eta(M)|_U$. If $\eta|_U$ is not exact, then $e^{\int_0^1 \gamma^*(\eta)}$ can have different values on different paths in U, and hence $\mathscr{F}_{e^\eta}(U) \approx \{0\}$. This shows that $\mathscr{F}_{e^\eta}$ is isomorphic to $\mathscr{F}_\eta(M)$, cf. Proposition 5.8 of [89].

Continuing with Exercise 6.F.7 of [80], we see that the Čech cohomology of M with coefficients in the sheaf $\mathscr{F}_\eta(M)$ is isomorphic to the singular cohomology of M with coefficients in the local system e^η, i.e

$$H^k(M; \mathscr{F}_\eta(M)) \approx H^k(M; e^\eta)$$

for all $k = 0, \ldots, m$. To summarize, we have the following isomorphisms between η-twisted Morse cohomology, Lichnerowicz cohomology, sheaf cohomology, and singular cohomology with coefficients in the local system e^η.

Theorem 5.28 *Let $f : M \to \mathbb{R}$ be a smooth Morse-Smale function on a closed finite dimensional smooth Riemannian manifold (M, g). For any $\eta \in \Omega^1_{cl}(M, \mathbb{R})$,*

$$H_k((C^*(f; e^\eta), \delta^\eta_*)) \approx H^k_{-\eta}(M) \approx H^k(M; \mathscr{F}_\eta(M)) \approx H^k(M; e^\eta)$$

for all $k = 0, \ldots, m$.

Remark 5.29 The techniques used in Chap. 4 can be used to prove directly that the η-twisted Morse cohomology groups are isomorphic to the singular cohomology groups with coefficients in e^η.

Chapter 6
Applications and Computations

In this chapter we demonstrate the effectiveness of twisted Morse complexes by collecting and expanding on some applications of homology and cohomology with local coefficients. We discuss relationships between parallel 1-forms and Lichnerowicz cohomology, and we show how to use twisted Morse cohomology to explicitly compute the Lichnerowicz cohomology of a surface. We also discuss Morse theoretic obstructions to a space being an associative H-space. Furthermore, we discuss Novikov homology and show how Novikov numbers can be computed using the results in Chaps. 4 and 5.

6.1 Parallel 1-Forms and Lichnerowicz Cohomology Computations

We begin this section by establishing some preliminary results that will be used to prove the following theorem of León, López, Marrero, and Padrón.

Theorem 6.1 (Theorem 4.5 of [25]) *Let M be a compact differentiable manifold of finite dimension m and η a closed 1-form on M with $\eta \neq 0$. Supposed that g is a Riemannian metric on M such that η is parallel with respect to g. Then, the cohomology $H_\eta^*(M)$ is trivial.*

Let η be a closed 1-form on a finite dimensional differentiable Riemannian manifold (M, g) of dimension m. Let U be the vector field dual to η under g, i.e. $\eta(X) = \mathsf{g}(X, U)$ for any vector field X on M, and for any $\xi \in \Omega^k(M, \mathbb{R})$ define

$$e(\eta)(\xi) = \eta \wedge \xi$$
$$d_\eta \xi = (d + e(\eta))\xi = d\xi + \eta \wedge \xi$$
$$\delta \xi = (-1)^{mk+m+1} \star d \star \xi$$

$$i_U(\xi) = (-1)^{mk+m} \star e(\eta) \star \xi = (-1)^{mk+m} \star (\eta \wedge \star \xi)$$
$$\delta_\eta \xi = (\delta + i_U)\xi,$$

where $\star$ is the Hodge star operator. Recall that the star isomorphism

$$\star : \Omega^k(M, \mathbb{R}) \to \Omega^{m-k}(M, \mathbb{R})$$

satisfies $\star \star \xi = (-1)^{k(m-k)} \xi$ for all $\xi \in \Omega^k(M, \mathbb{R})$, and note that the sign $(-1)^{k(m-k)}$ depends on the degree of the form ξ. Moreover, δ is the adjoint of d with respect to the inner product on differential forms given by

$$<\alpha, \beta> = \int_M \alpha \wedge \star \beta.$$

That is,

$$<d\alpha, \beta> = <\alpha, \delta \beta>$$

for all $\alpha \in \Omega^{k-1}(M, \mathbb{R})$ and $\beta \in \Omega^k(M, \mathbb{R})$, cf. Proposition 6.2 of [89].

Proposition 6.2 *The adjoint of $e(\eta)$ is i_U.*

Proof Let $\alpha \in \Omega^{k-1}(M, \mathbb{R})$ and $\beta \in \Omega^k(M, \mathbb{R})$.

$$\alpha \wedge \star i_U \beta = \alpha \wedge \star (-1)^{mk+m} \star e(\eta) \star \beta$$
$$= (-1)^{mk+m} \alpha \wedge \star \star (\eta \wedge \star \beta)$$
$$= (-1)^{mk+m}(-1)^{(m-k+1)(k-1)} \alpha \wedge \eta \wedge \star \beta$$
$$= (-1)^{k^2+1}(-1)^{k-1} \eta \wedge \alpha \wedge \star \beta$$
$$= (-1)^{k(k+1)} e(\eta) \alpha \wedge \star \beta$$

Therefore,

$$<e(\eta)\alpha, \beta> = <\alpha, i_U \beta>.$$

$\square$

Corollary 6.3 *The adjoint of d_η is δ_η.*

The notation i_U is often used to denote interior product by the vector field U. That is, if $\alpha \in \Lambda^k(T^*M)$ and $X \in \Lambda^{k-1}(T_*M)$, then

$$(i_U \alpha, X) = (\alpha, U \wedge X),$$

6.1 Parallel 1-Forms and Lichnerowicz Cohomology Computations

where $(\alpha, U \wedge X)$ denotes the pairing of a k-covector with a k-vector, cf. Definitions 2.9 and 2.11 of [89].

Lemma 6.4 *The formula for i_U given above in terms of $\star$ and $e(\eta)$ coincides with the interior product by U.*

Proof An inner product $<,>$ on a real vector space V induces an inner product on V^* via the isomorphism given by the inner product. An inner product on $\Lambda^k(V^*)$ is then determined by the Gram determinant, cf. equation (3.3.1) of [45]. The inner product on $\Lambda^k(V^*)$ can also be computed using the star operator:

$$<\alpha, \beta> = \star(\alpha \wedge \star\beta),$$

cf. Lemma 3.3.2 of [45]. So, if U is dual to the 1-form η and α is a 1-form with dual vector field A, we have

$$i_U(\alpha) = (-1)^{2m} \star (\eta \wedge \star\alpha)$$
$$= <\eta, \alpha>$$
$$= <U, A>$$
$$= \alpha(U).$$

The result now follows from the fact that both i_U and the interior product are anti-derivations, cf. Lemma 3.3.2 of [34] and Proposition 2.12 of [89]. □

The previous lemma implies that Cartan's formula $\mathcal{L}_U = i_U \circ d + d \circ i_U$ for the Lie derivative on $\Omega^*(M, \mathbb{R})$ holds with respect to the definition of i_U given above, cf. Lemma 3.5.1 of [34] or Proposition 2.25 of [89].

Proposition 6.5 *The adjoint of $\mathcal{L}_U$ is $\delta \circ e(\eta) + e(\eta) \circ \delta$.*

Proof We have the following for all $\alpha, \beta \in \Omega^k(M, \mathbb{R})$.

$$<\mathcal{L}_U\alpha, \beta> = <(i_U \circ d + d \circ i_U)\alpha, \beta>$$
$$= <i_U(d\alpha), \beta> + <di_U(\alpha), \beta>$$
$$= <\alpha, \delta e(\eta)(\beta)> + <\alpha, e(\eta)(\delta\beta)>$$
$$= <\alpha, (\delta \circ e(\eta) + e(\eta) \circ \delta)\beta>$$

□

Lemma 6.6 *If $\eta(U) = 1$, then $i_U \circ e(\eta) + e(\eta) \circ i_U = Id$.*

Proof The identity holds on $\Omega^0(M, \mathbb{R})$ because $i_U(f) = 0$ for all $f \in \Omega^0(M, \mathbb{R})$. To show that the identity holds on $\Omega^k(M, \mathbb{R})$ for $k \geq 1$, it suffices to show that it holds on elements of the form $\alpha_1 \wedge \cdots \wedge \alpha_k \in \Omega^k(M, \mathbb{R})$, where $\alpha_j \in \Omega^1(M, \mathbb{R})$ for all $j = 1, \ldots, k$.

Using the fact that i_U is an anti-derivation, cf. Proposition 2.12 of [89], the first term $i_U \circ e(\eta)$ evaluated on $\alpha_1 \wedge \cdots \wedge \alpha_k$ is

$$i_U(\eta \wedge \alpha_1 \wedge \cdots \wedge \alpha_k) = \eta(U)\alpha_1 \wedge \cdots \wedge \alpha_k + \sum_{j=1}^{k}(-1)^j \alpha_j(U)\,\eta \wedge \alpha_1 \wedge \cdots \wedge \widehat{\alpha_j} \wedge \cdots \wedge \alpha_k,$$

and the second term $e(\eta) \circ i_U$ evaluated on $\alpha_1 \wedge \cdots \wedge \alpha_k$ is

$$\eta \wedge i_U(\alpha_1 \wedge \cdots \wedge \alpha_k) = \eta \wedge \sum_{j=1}^{k}(-1)^{j+1}\alpha_j(U)\alpha_1 \wedge \cdots \wedge \widehat{\alpha_j} \wedge \cdots \wedge \alpha_k.$$

These terms sum to $\eta(U)\alpha_1 \wedge \cdots \wedge \alpha_k$, where $\eta(U) = 1$. □

Proof of Theorem 6.1 If η is parallel and non-zero it has constant length, and we may assume without loss of generality that $\|\eta\| = 1$, i.e. $\eta(U) = 1$. In addition, the star operator commutes with the Lie derivative $\mathcal{L}_U$ since U is Killing, i.e. $\star \mathcal{L}_U = \mathcal{L}_U \star$, and $\mathcal{L}_U$ is anti self-adjoint, cf. (3.7.8) of [34]. Therefore, for all $\xi \in \Omega^k(M, \mathbb{R})$

$$<\mathcal{L}_U \xi, \xi> = <\xi, -\mathcal{L}_U \xi> = -<\mathcal{L}_U \xi, \xi> = 0.$$

Furthermore,

$$\mathcal{L}_U = i_U \circ d + d \circ i_U = -\delta \circ e(\eta) - e(\eta) \circ \delta$$

by Cartan's formula and Proposition 6.5, which implies that d, δ, i_U, $e(\eta)$, d_η, and δ_η all commute with $\mathcal{L}_U$. Moreover, $\eta(U) = 1$ implies that

$$i_U \circ d_\eta + d_\eta \circ i_U = i_U \circ d + i_U \circ e(\eta) + d \circ i_U + e(\eta) \circ i_U$$
$$= \mathcal{L}_U + i_U \circ e(\eta) + e(\eta) \circ i_U$$
$$= \mathcal{L}_U + Id.$$

Hence,

$$\mathcal{L}_U = i_U \circ d_\eta + d_\eta \circ i_U - Id.$$

Now, the operators d and d_η have the same symbol for any closed 1-form η, and thus $(\Omega^*(M, \mathbb{R}), d_\eta)$ is elliptic, cf. Proposition 4.4 of [25]. Hence, there is a Hodge decomposition

$$\Omega^k(M, \mathbb{R}) = \mathcal{H}_\eta^k(M) \oplus d_\eta(\Omega^{k-1}(M)) \oplus \delta_\eta(\Omega^{k+1}(M))$$

where

$$\mathcal{H}_\eta^k(M) = \{\xi \in \Omega^k(M, \mathbb{R})|\ d_\eta \xi = 0 \text{ and } \delta_\eta \xi = 0\}.$$

6.1 Parallel 1-Forms and Lichnerowicz Cohomology Computations

If $\xi \in \mathcal{H}_\eta^k(M)$, then $\mathcal{L}_U \xi \in \mathcal{H}_\eta^k(M)$ because

$$d_\eta \mathcal{L}_U \xi = \mathcal{L}_U d_\eta \xi = 0 \quad \text{and} \quad \delta_\eta \mathcal{L}_U \xi = \mathcal{L}_U \delta_\eta \xi = 0.$$

Moreover, the above formula for $\mathcal{L}_U$ says that $\mathcal{L}_U \xi = d_\eta i_U(\xi) - \xi$, and thus the Hodge decomposition implies that $\mathcal{L}_U \xi = -\xi$. Hence,

$$<\xi,\xi> = <\xi,-\mathcal{L}_U \xi> = 0,$$

and $\mathcal{H}_\eta^k(M) \approx 0$ for all k. □

Combining the η-Twisted Morse de Rham Theorem (Theorem 5.22) with Theorem 6.1 gives the following.

Theorem 6.7 *Let $f : M \to \mathbb{R}$ be a smooth Morse-Smale function on a closed finite dimensional smooth Riemannian manifold (M, g). If η is a nonzero closed 1-form on M that is parallel with respect to g, then*

$$H_k((C^*(f; e^\eta), \delta_*^\eta)) \approx 0$$

for all k.

Proof The 1-form η is parallel with respect to g if any only if $-\eta$ is parallel with respect to g. Thus, Theorems 5.22 and 6.1, imply that

$$H_k((C^*(f; e^\eta), \delta_*^\eta)) \approx H_{-\eta}^k(M) = 0$$

for all k. □

Corollary 6.8 (Parallel 1-Form Obstruction) *Let $f : M \to \mathbb{R}$ be a smooth Morse-Smale function on a closed finite dimensional smooth Riemannian manifold M, and assume there exists a nonzero closed 1-form η on M such that $H_k((C^*(f; e^\eta), \delta_*^\eta)) \neq 0$ for some k. Then for any nonzero closed 1-form ζ on M such that $[\zeta] = [\eta] \in H^1(M; \mathbb{R})$ the 1-form ζ is not parallel with respect to any Riemannian metric on M.*

Proof By Propositions 2.6 and 5.4 (or Theorems 5.14 and 5.22) we have

$$H_k((C^*(f; e^\eta), \delta_*^\eta)) \neq 0 \quad \Rightarrow \quad H_k((C^*(f; e^\zeta), \delta_*^\zeta)) \neq 0.$$

By Theorem 5.22 these homology groups are independent of the Riemannian metric used to define them, and hence the result follows from Theorem 6.7. □

Corollary 6.9 (Euler Number Parallel 1-Form Obstruction) *If M is a closed finite dimensional smooth manifold with nonzero Euler number, then there are no nonzero closed parallel 1-forms on M with respect to any Riemannian metric on M.*

Proof Corollary 5.25 implies that for any nonzero closed 1-form $\eta \in \Omega_{cl}(M, \mathbb{R})$, $\mathcal{X}^\eta(M) = \mathcal{X}(M) \neq 0$. Therefore, $H_\eta^*(M) \neq 0$, and Theorem 6.1 implies that η is not parallel with respect to any Riemannian metric on M. □

Example 6.10 (The Lichnerowicz Cohomology of a Cylinder) Let (N, g) be a compact smooth Riemannian manifold and let $M = N \times S^1$. Define $\eta = \pi_2^*(d\theta)$ to be the 1-form on M given by pulling back the closed 1-form $d\theta$ from Example 2.14 under the projection $\pi_2 : M \times S^1 \to S^1$ onto the second component. The 1-form η is parallel on M with respect to the Riemannian metric

$$\tilde{g} = \pi_1^*(g) + \eta \otimes \eta,$$

where $\pi_1 : M \times S^1 \to M$ denotes projection onto the first factor. Therefore, by Theorem 6.7 we see that for any Morse-Smale function $f : M \to \mathbb{R}$ the η-twisted Morse cohomology groups

$$H_k((C^*(f; e^\eta), \delta_*^\eta)) \approx 0$$

for all k.

Now assume in addition that (N, g, λ) is a contact manifold, and define

$$\Omega = \pi_1^*(d\lambda) + \pi_2^*(d\theta) \wedge \pi_1^*(\lambda).$$

Then Ω is a nondegenerate 2-form on M such that

$$d\Omega = \pi_1^*(d^2\lambda) + \pi_2^*(d(d\theta)) \wedge \pi_1^*(\lambda) - \pi_2^*(d\theta) \wedge \pi_1^*(d\lambda)$$
$$= -\pi_2^*(d\theta) \wedge \pi_1^*(d\lambda)$$
$$= -\pi_2^*(d\theta) \wedge \Omega.$$

Hence, $M = N \times S^1$ is a locally conformal symplectic (LCS) manifold with Lee form $\eta = \pi_2^*(d\theta)$, cf. Proposition 5.10. Since η is parallel with respect to the Riemannian metric $\tilde{g}$ on M, this gives a class of LCS manifolds (M, Ω) with Lee form η such that the η-twisted Morse cohomology groups $H_k((C^*(f; e^\eta), \delta_*^\eta))$ are zero for all k, cf. Example 4.10(2) of [25].

Remark 6.11 In the previous example the η-twisted Euler number

$$\mathcal{X}^\eta(M) = \sum_{k=0}^m (-1)^k \dim H_\eta^k(M) = 0.$$

Thus, Corollary 5.25 implies that the untwisted Euler number $\mathcal{X}(M)$ is also zero. The fact that $\mathcal{X}(N \times S^1) = 0$ can also be seen directly from the Künneth Theorem, cf. Theorem VI.3.2 of [17], or the fact that $\mathcal{X}(N \times S^1) = \mathcal{X}(N)\mathcal{X}(S^1)$.

6.1 Parallel 1-Forms and Lichnerowicz Cohomology Computations

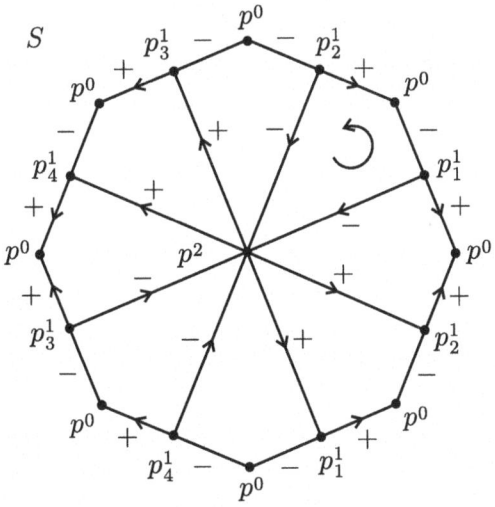

A surface of genus 2

Example 6.12 (The Lichnerowicz Cohomology of a Surface of Genus Two)
Consider a surface S of genus 2 as the connected sum of two tori. If we view each tori as a square whose opposite sides are identified, then the surface of genus 2 can be viewed as an octagon where every other edge on each half of the octagon is identified, cf. Section I.5 of [51]. By applying the construction used in the proof of Theorem 4.12 to the octagon representing S we can construct a Morse-Smale pair (f, g) with one critical point of index 0, four critical points of index 1, and one critical point of index 2. The critical point p^0 of index 0 will correspond to the vertices of the octagon, which are all identified, the four critical points $p_1^1, p_2^1, p_3^1, p_4^1$, of index 1 will be on the edges of the octagon, which are identified in pairs, and the critical point p^2 of index 2 will be on the interior of the octagon.

As shown in the diagram, there are exactly two gradient flow lines between each pair of critical points of relative index one, and they have opposite signs. Hence, the (untwisted) coboundary operators δ_* are all zero in the Morse-Smale-Witten cochain complex.

$$0 \longrightarrow C^0(f; \mathbb{R}) \xrightarrow{\delta_0} C^1(f; \mathbb{R}) \xrightarrow{\delta_1} C^2(f; \mathbb{R}) \xrightarrow{\delta_2} 0$$

Therefore,

$$H_k((C^*(f;\mathbb{R}), \delta_*)) \approx \begin{cases} \mathbb{R} & \text{if } k = 0 \\ \mathbb{R} \oplus \mathbb{R} \oplus \mathbb{R} \oplus \mathbb{R} & \text{if } k = 1 \\ \mathbb{R} & \text{if } k = 2, \end{cases}$$

and Theorems 5.14 and 5.22 imply that $H^1_{\mathrm{dR}}(S; \mathbb{R}) \approx \mathbb{R}^4$. Moreover, there are smooth paths γ_i, such that $\mathrm{Im}(\gamma_i) = \overline{W^u(p_i^1)}$ for $i = 1, 2, 3, 4$, that represent a basis for $H_1(S; \mathbb{R})$. Hence, $H^1_{\mathrm{dR}}(S; \mathbb{R})$ has a basis represented by closed 1-forms $\eta_j \in \Omega_{\mathrm{cl}}(M, \mathbb{R})$ for $j = 1, 2, 3, 4$ such that

$$\int_{\gamma_i} \eta_j = \begin{cases} 1 \text{ if } i = j \\ 0 \text{ if } i \neq j, \end{cases}$$

since the identification of $H^1_{\mathrm{dR}}(S; \mathbb{R})$ with $\mathrm{Hom}(H_1(S; \mathbb{R}), \mathbb{R})$ is given by integrating a representative of the de Rham cohomology class over a smooth representative of the homology class, cf. Sections V.5 and V.7 of [17].

To compute $H_k((C^*(f; e^{\eta_j}), \delta_*^{\eta_j})) \approx H^k_{-\eta_j}(S)$ we need to consider the integrals of η_j over the gradient flow lines from p^2 to the critical points of index 1. To address this we apply Stokes' Theorem for smooth singular chains, cf. Theorem 8.4 of [82] or Theorem 4.7 of [89]. For instance, consider the integral of η_1 over the boundary of a smooth singular 2-cube $\sigma : [0, 1] \times [0, 1] \to S$ whose four faces map injectively onto the following smoothly embedded loops and closed intervals in the counterclockwise direction shown in the diagram.

1. The loop $\overline{W^u(p_1^1)}$.
2. The closure of the image of the positive gradient flow line from p_2^1 to p^0.
3. The loop $\overline{W(p^2, p_2^1)}$.
4. The closure of the image of the positive gradient flow line from p_2^1 to p^0.

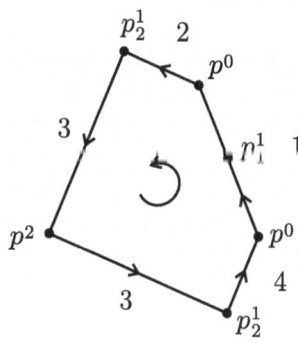

Since

$$\int_{\partial \sigma} \eta_1 = \int_\sigma d\eta_1 = 0,$$

and the integrals over the closure of the image of the positive gradient flow line from p_2^1 to p^0 cancel out, we see that the integrals of η_1 over the loops $\overline{W^u(p_1^1)}$ and $\overline{W(p^2, p_2^1)}$ are opposite each other. Similarly, the integrals of η_1 over the loops

6.1 Parallel 1-Forms and Lichnerowicz Cohomology Computations

$\overline{W(p^2, p_1^1)}$, $\overline{W(p^2, p_3^1)}$, and $\overline{W(p^2, p_4^1)}$ are all zero, since the integrals of η_1 over the loops $\overline{W^u(p_2^1)}$, $\overline{W^u(p_4^1)}$, and $\overline{W^u(p_3^1)}$ are zero. This gives enough information to compute $H_k((C^*(f; e^{\eta_1}), \delta_*^{\eta_1})) \approx H_{-\eta_1}^k(S)$ for all k.

$$0 \longrightarrow C^0(f; e^{\eta_1}) \xrightarrow{\delta_0^{\eta_1}} C^1(f; e^{\eta_1}) \xrightarrow{\delta_1^{\eta_1}} C^2(f; e^{\eta_1}) \xrightarrow{\delta_2^{\eta_1}} 0$$

To see this, let $\theta \in C^0(f; e^{\eta_1})$, and recall from Definition 5.6 that

$$(\delta_0^{\eta_1}\theta)(p_i^1) = \sum_{v \in M(p_i^1, p^0)} \epsilon(v) \exp\left(\int_{\mathbb{R}} (\gamma^v)^*(\eta_1)\right) \theta(p^0) \in e_{p_i^1}^{\eta_1},$$

for any critical point $p_i^1 \in Cr_1(f)$, where $\gamma^v : \mathbb{R} \to M$ is any continuous path from p^0 to p_i^1 whose image coincides with the image of $v \in M(p_i^1, p^0)$ and $\epsilon(v) = \pm 1$ is the sign determined by the orientation on $M(p_i^1, p^0)$. For $i \neq 1$, the integral of η_1 over the loop $\overline{W^u(p_i^1)}$ is zero, and hence the two integrals of η_1 in the above sum are the same. Thus, $(\delta_0^{\eta_1}\theta)(p_i^1) = 0$ for $i \neq 1$, since the two gradient flow lines from p_i^1 to p^0 have opposite signs. However, $(\delta_0^{\eta_1}\theta)(p_1^1) \neq 0$ for $\theta \neq 0$ because the two integrals have different values. Specifically, if the integral of η_1 along the positive gradient flow line is a, then the integral of η_1 along the negative gradient flow line will be $a \pm 1$, where the sign is determined by whatever orientation is chosen for the generator γ_1 of $H_1(S; \mathbb{R})$. Therefore,

$$(\delta_0^{\eta_1}\theta)(p_1^1) = (e^a - e^{a \pm 1})\theta(p^0) \neq 0$$

if $\theta(p^0) \neq 0$, and

$$H_0((C^*(f; e^{\eta_1}), \delta_*^{\eta_1})) \approx 0.$$

Moreover, if θ_j is a basis for $C^1(f; e^{\eta_1})$ such that

$$\theta_j(p_i^1) = \begin{cases} 1 \text{ if } i = j \\ 0 \text{ if } i \neq j \end{cases}$$

for $j = 1, 2, 3, 4$, then the above argument shows that θ_1 generates Im $\delta_0^{\eta_1}$.

Now consider $\delta_1^{\eta_1}(\theta_j)$ for $j = 1, 2, 3, 4$. Using the notation from Definition 5.6 we have

$$(\delta_1^{\eta_1}\theta_j)(p^2) = \sum_{p_i^1 \in Cr_1(f)} \sum_{v \in M(p^2, p_i^1)} \epsilon(v) \exp\left(\int_{\mathbb{R}} (\gamma^v)^*(\eta_1)\right) \theta_j(p_i^1)$$

$$= \sum_{v \in M(p^2, p_j^1)} \epsilon(v) \exp\left(\int_{\mathbb{R}} (\gamma^v)^*(\eta_1)\right) 1 \in e_{p^2}^{\eta_1}.$$

Since the integrals of η_1 over the loops $\overline{W(p^2, p_1^1)}$, $\overline{W(p^2, p_3^1)}$, and $\overline{W(p^2, p_4^1)}$ are zero and the two gradient flow lines from p^2 to p_i^1 have opposite signs, we see that $\theta_1, \theta_3, \theta_4 \in \ker \delta_1^{\eta_1}$. However, the integral of η_1 over the loop $\overline{W(p^2, p_2^1)}$ is ± 1, where the sign is determined by whatever orientation is chosen for the generator γ_1 of $H_1(S; \mathbb{R})$. Therefore, $\delta_1^{\eta_1}(\theta_2) \neq 0$, and we have

$$H^k_{-\eta_1}(S) \approx H_k((C^*(f; e^{\eta_1}), \delta_*^{\eta_1})) \approx \begin{cases} 0 & \text{if } k = 0 \\ \mathbb{R} \oplus \mathbb{R} & \text{if } k = 1 \\ 0 & \text{if } k = 2. \end{cases}$$

It's clear that the Lichnerowicz cohomology $H^k_{-\eta_j}(S) \approx H_k((C^*(f; e^{\eta_j}), \delta_*^{\eta_j}))$ is the same for $j = 2, 3, 4$. The Lichnerowicz cohomology $H^k_{-\eta}(S)$ for a general closed 1-form $\eta \in \Omega^1_{cl}(S, \mathbb{R})$ can be computed using the invariance of the η-twisted Euler number (Corollary 5.25), i.e. $\mathcal{X}^\eta(S) = \mathcal{X}(S) = -2$, and the fact that $\{[\eta_1], [\eta_2], [\eta_3], [\eta_4]\}$ is a basis for $H^1_{dR}(S; \mathbb{R})$. That is, Corollary 5.25 and the placement of the 6 generators of $(C^*(f; e^\eta), \delta_*^\eta)$ imply that there are only 4 possibilities for $H^*_{-\eta}(S)$. We claim for any $\eta \in \Omega^1_{cl}(S, \mathbb{R})$ with $[\eta] \neq 0$, $H^*_{-\eta}(S)$ is the same as $H^*_{-\eta_1}(S)$.

To see this, let $\eta = a_1 \eta_1 + a_2 \eta_2 + a_3 \eta_3 + a_4 \eta_4$ for some $a_1, a_2, a_3, a_4 \in \mathbb{R}$ and assume that $a_j \neq 0$ for some j, i.e. η is not exact. Letting γ^+ and γ^- be paths from p^0 to p_j^1 parameterizing the elements of $M(p_j^1, p^0)$ with signs $+$ and $-$ respectively we compute as follows. For any $\theta \in C^0(f; e^\eta)$, we have $(\delta_0^\eta \theta)(p_j^1) = $

$$\sum_{v \in M(p_j^1, p^0)} \epsilon(v) \exp\left(\int_{\mathbb{R}} (\gamma^v)^*(\eta)\right) \theta(p^0)$$

$$= \left(e^{\int_{\gamma^+} \eta} - e^{\int_{\gamma^-} \eta}\right) \theta(p^0)$$

$$= \left(e^{a_1 \int_{\gamma^+} \eta_1} e^{a_2 \int_{\gamma^+} \eta_2} e^{a_3 \int_{\gamma^+} \eta_3} e^{a_4 \int_{\gamma^+} \eta_4}\right.$$

$$\left. - e^{a_1 \int_{\gamma^-} \eta_1} e^{a_2 \int_{\gamma^-} \eta_2} e^{a_3 \int_{\gamma^-} \eta_3} e^{a_4 \int_{\gamma^-} \eta_4}\right) \theta(p^0)$$

$$= e^A e^B e^C (e^{a_j a} - e^{a_j(a \pm 1)}) \theta(p^0)$$

for some $a, A, B, C \in \mathbb{R}$, because

$$\int_{\gamma^-} \eta_i = \int_{\gamma^+} \eta_i \text{ if } i \neq j \quad \text{and} \quad \int_{\gamma^-} \eta_j = a \pm 1 \text{ if } \int_{\gamma^+} \eta_j = a.$$

6.1 Parallel 1-Forms and Lichnerowicz Cohomology Computations

This shows that $H_0((C^*(f;e^\eta),\delta_*^\eta)) \approx 0$.

Now consider the basis element $\theta_l \in C^1(f;e^\eta)$, where $\overline{W(p^2,p_l^1)}$ is on the opposite side of the singular 2-cube from $\overline{W^u(p_j^1)}$, and let γ^+ and γ^- be paths from p_l^1 to p^2 parameterizing the elements of $\mathcal{M}(p^2,p_l^1)$ with signs $+$ and $-$ respectively.

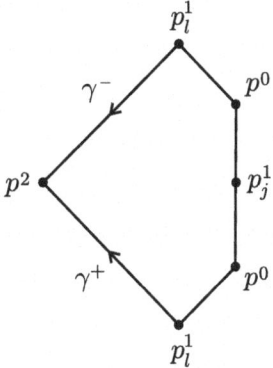

$(\delta_1^\eta \theta_l)(p^2) =$

$$\sum_{p_i^1 \in Cr_1(f)} \sum_{v \in \mathcal{M}(p^2,p_i^1)} \epsilon(v) \exp\left(\int_\mathbb{R} (\gamma^v)^*(\eta)\right) \theta_l(p_i^1)$$

$$= \sum_{v \in \mathcal{M}(p^2,p_l^1)} \epsilon(v) \exp\left(\int_\mathbb{R} (\gamma^v)^*(\eta)\right) 1$$

$$= \left(e^{a_1 \int_{\gamma^+} \eta_1} e^{a_2 \int_{\gamma^+} \eta_2} e^{a_3 \int_{\gamma^+} \eta_3} e^{a_4 \int_{\gamma^+} \eta_4} - e^{a_1 \int_{\gamma^-} \eta_1} e^{a_2 \int_{\gamma^-} \eta_2} e^{a_3 \int_{\gamma^-} \eta_3} e^{a_4 \int_{\gamma^-} \eta_4}\right)$$

$$= e^A e^B e^C (e^{a_j a} - e^{a_j(a \pm 1)})$$

for some $a, A, B, C \in \mathbb{R}$, because

$$\int_{\gamma^-} \eta_i = \int_{\gamma^+} \eta_i \text{ if } i \neq j \quad \text{and} \quad \int_{\gamma^-} \eta_j = a \pm 1 \text{ if } \int_{\gamma^+} \eta_j = a.$$

Thus $\delta_1^\eta \theta_l \neq 0$ when $a_j \neq 0$, and $H_2((C^*(f;e^\eta),\delta_*^\eta)) \approx 0$. Therefore,

$$H_{-\eta}^k(S) \approx H_k((C^*(f;e^\eta),\delta_*^\eta)) \approx \begin{cases} 0 & \text{if } k = 0 \\ \mathbb{R} \oplus \mathbb{R} & \text{if } k = 1 \\ 0 & \text{if } k = 2 \end{cases}$$

for any $\eta \in \Omega_{cl}^1(S,\mathbb{R})$ with $[\eta] \neq 0$, because $\mathcal{X}^\eta(S) = -2$.

6.2 H-Spaces

Recall that an H-space is a topological space X together with a continuous map $m : X \times X \to X$ and an element $e \in X$ such that $m(e, \cdot) : X \to X$ and $m(\cdot, e) : X \to X$ are homotopic to the identity through maps that preserve e. The map m is called **multiplication** and the element e is called the **homotopy unit**. This definition can be weakened by removing the assumption that the homotopies preserve e or strengthened by requiring that e be a strict identity, and all three definitions are equivalent for CW-complexes, cf. Section 3.C of [36]. Standard examples of H-spaces include the based loop space of a topological space, where the multiplication map is given by concatenation, and topological groups, i.e. topological spaces with a group structure such that both the multiplication and inverse maps are continuous. For both of these examples the multiplication map m is associative.

We will only consider H-spaces where m is associative up to homotopy because we are interested in the *Pontryagin ring*. For any commutative ring R with unit the multiplication map m on an H-space X induces what is known as the **Pontryagin product** on the homology:

$$H_*(X; R) \otimes H_*(X; R) \xrightarrow{\times} H_*(X \times X; R) \xrightarrow{m_*} H_*(X; R),$$

where the first map is the homology cross product. The assumption that m is associative up to homotopy implies that the Pontryagin product is associative, and hence $H_*(X; R)$ is a graded ring with unit known as the **Pontryagin ring**, cf. Section VII.3 of [27] or Section III.7 of [93].

Recently, Albers et al. extended the Pontryagin product to homology with local coefficients in a system of rank one R-modules $\mathcal{L}$ on a path connected space X, where R is a commutative ring with unit [2]. That is, $\mathcal{L}$ is a bundle of abelian groups such that each fiber has the structure of a free rank one R-module and the homomorphisms γ_* are all R-module isomorphisms, i.e. γ_* is given by multiplication by an element of the group of invertible elements $R^\times \subseteq R$. Any bundle of abelian groups with fiber $\mathbb{Z}$ is an example of a local coefficient system of rank one $\mathbb{Z}$-modules, and e^η, where η is a closed smooth 1-form on a smooth manifold, is an example of a local coefficient system of rank one $\mathbb{R}$-modules (see Remark 2.7).

Proposition 6.13 (Proposition 1 [2]) *Let R be a commutative ring with unit and X a path-connected associative H-space. For any local coefficient system of rank one R-modules $\mathcal{L}$ on X, the multiplication $m : X \times X \to X$ induces a unital ring structure on $H_*(X; \mathcal{L})$. The unit is represented by the class of a point.*

The ring structure is defined at the chain level as the composition

$$C_*(X; \mathcal{L}) \otimes C_*(X; \mathcal{L}) \xrightarrow{B} C_*(X \times X; pr_1^*\mathcal{L} \otimes_R pr_2^*\mathcal{L}) \xrightarrow{m_*} C_*(X; \mathcal{L}),$$

6.2 H-Spaces

where B is the Eilenberg-MacLane shuffle map with local coefficients in $\mathcal{L}$, $pr_i : X \times X \to X$ is a projection map for $i = 1, 2$, and m_* is induced by the multiplication m. The fact that $\mathcal{L}$ is a local system of *rank one* R modules is central to the proof of Proposition 6.13, since this implies that there is an isomorphism

$$pr_1^* \mathcal{L} \otimes_R pr_2^* \mathcal{L} \cong m^* \mathcal{L}.$$

This fact is proved in Lemma 1 of [2] using the natural isomorphism $R \otimes_R R \cong R$ given by multiplication in R.

Albers et al. used Proposition 6.13 to prove a vanishing result for the homology of an H-space with local coefficients in certain systems of rank one R-modules (Proposition 3 of [2]).

Proposition 6.14 *Let R be a commutative ring with unit and $\mathcal{L}$ a local coefficient system of rank one R-modules on a path connected associative H-space X with homotopy unit $e \in X$. Assume there exists $[\gamma] \in \pi_1(X, e)$ such that*

1. *$\gamma_*(s) = gs$ for some $g \in R_e^\times$ with $g \neq 1$, i.e. $\gamma_* \neq id$, and*
2. *$g^{-1} - 1 \in R_e^\times$, i.e. $g^{-1} - 1$ is invertible.*

Then $H_*(X; \mathcal{L}) = 0$.

Proof If $e \in X$ is the homotopy unit, then Proposition 6.13 says that $[1 \cdot e] \in H_0(X; \mathcal{L})$ is the unit in the Pontryagin ring $H_*(X; \mathcal{L})$, where $1 \in R_e$ (see Sect. 4.1). Thus, it suffices to show that (1) and (2) imply that $[1 \cdot e] = [0] \in H_*(X; \mathcal{L})$.

If $\gamma : [0, 1] \to X$ is a loop based at e, then $\gamma_*(s) = gs$ for some $g \in R_e^\times$, and for any $s \in R_e$ we have $s \cdot \gamma \in C_1(X; \mathcal{L})$ with

$$\partial_1(s \cdot \gamma) = \gamma_*^{-1}(s) \cdot e - s \cdot e = \left(g^{-1}s - s\right) \cdot e$$

(see Definition 4.2). In particular, for $s = 1 \in R_e$ we have $\partial_1(1 \cdot \gamma) = \left(g^{-1} - 1\right) \cdot e$. Hence, if $[\gamma] \in \pi_1(M, e)$ satisfies conditions (1) and (2), we can set $s = \left(g^{-1} - 1\right)^{-1}$ in the above equation for $\partial_1(s \cdot \gamma)$ and we have

$$\partial_1\left((g^{-1} - 1)^{-1} \cdot \gamma\right) = \left(g^{-1}(g^{-1} - 1)^{-1} - (g^{-1} - 1)^{-1}\right) \cdot e$$
$$= \left((g^{-1} - 1)(g^{-1} - 1)^{-1}\right) \cdot e$$
$$= 1 \cdot e.$$

Therefore, $[1 \cdot e] = [0] \in H_*(X; \mathcal{L})$. □

Note If $R = \mathbb{Z}$, then conditions (1) and (2) cannot hold simultaneously because $\gamma_*(s) = \pm s$ when $R_e = \mathbb{Z}$ and $-2 \notin \mathbb{Z}^\times$. Similarly, if $R = \mathbb{Z}_2$, then condition (1) cannot hold. However, if R is a field and $\mathcal{L}$ is not simple, then there always exists some $[\gamma] \in \pi_1(X, e)$ such that (1) and (2) hold. This observation leads to the following obstruction to the existence of an associative H-space structure.

Corollary 6.15 *If X is a path connected topological space and there exists a local coefficient system $\mathcal{L}$ of rank one vector spaces on X, i.e. a line bundle over some field, such that*

1. *$\mathcal{L}$ is not simple, i.e. $\mathcal{L}$ is not isomorphic to a constant bundle, and*
2. *$H_*(X; \mathcal{L}) \neq 0$,*

then X is not an associative H-space.

We now interpret and apply the above results within the context of twisted Morse homology.

Theorem 6.16 *Let $f : M \to \mathbb{R}$ be a smooth Morse-Smale function on a closed path connected finite dimensional smooth Riemannian manifold (M, g), and let $\mathcal{L}$ be a local coefficient system of rank one R-modules on M, where R is a commutative ring with unit. Assume that M is an associative H-space with homotopy unit $e \in M$, and there exists $[\gamma] \in \pi_1(X, e)$ such that*

1. *$\gamma_*(s) = gs$ for some $g \in R_e^\times$ with $g \neq 1$, i.e. $\gamma_* \neq id$, and*
2. *$g^{-1} - 1 \in R_e^\times$, i.e. $g^{-1} - 1$ is invertible.*

Then $H_k((C_*(f; \mathcal{L}), \partial_*^\mathcal{L})) = 0$ for all $k = 0, \ldots, m$.

Proof Note that if we "forget" the product structure, then $\mathcal{L}$ is a bundle of abelian groups. So, the twisted Morse-Smale-Witten chain complex $(C_*(f; \mathcal{L}), \partial_*^\mathcal{L})$ is defined, and by the Twisted Morse Homology Theorem (Theorem 4.1) the homology groups of $(C_*(f; \mathcal{L}), \partial_*^\mathcal{L})$ are isomorphic to the singular homology groups of M with coefficients in the bundle of abelian groups $\mathcal{L}$. Moreover, the singular homology groups of M with coefficients in the bundle of abelian groups obtained by "forgetting" the product structure on $\mathcal{L}$ are isomorphic to the singular homology groups of M with coefficients in the R-module $\mathcal{L}$. The result now follows from Proposition 6.14. □

Corollary 6.17 (Associative H-Space Obstruction) *Let $f : M \to \mathbb{R}$ be a smooth Morse-Smale function on a closed path connected finite dimensional smooth Riemannian manifold (M, g). If there exists a local coefficient system $\mathcal{L}$ of rank one vector spaces on M, i.e. a line bundle over some field, such that*

1. *$\mathcal{L}$ is not simple, i.e. $\mathcal{L}$ is not isomorphic to a constant bundle, and*
2. *$H_k((C_*(f; \mathcal{L}), \partial_*^\mathcal{L})) \neq 0$ for some k,*

then M is not an associative H-space.

Corollary 6.18 (Euler Number Associative H-Space Obstruction) *If M is a closed path connected finite dimensional smooth manifold with $H_{dR}^1(M; \mathbb{R}) \neq 0$ and $\chi(M) \neq 0$, then M is not an associative H-space.*

Proof Pick a closed 1-form $\eta \in \Omega_{cl}(M, \mathbb{R})$ such that the de Rham cohomology class $[\eta] \neq 0$. Then e^η is a non-simple local coefficient system of rank one vector spaces on M. Moreover, by Theorem 4.12 and Corollary 4.22 we have that $H_*(M; e^\eta) \neq 0$, since $\mathcal{X}_{e^\eta}(M) = \mathcal{X}(M) \neq 0$. The result now follows from Corollary 6.15. □

Note Since both based loop spaces and topological groups are associative H-spaces, Corollaries 6.17 and 6.18 give obstructions to closed finite dimensional smooth manifolds being based loop spaces or topological groups, cf. [12].

Example 6.19 (The Twisted Homology of a Circle) The manifold S^1 is a topological group, and hence an associative H-space. Thus, Theorem 6.16 implies that the twisted Morse homology of S^1 vanishes for any non-simple local coefficient system $\mathcal{L}$ of rank one vector spaces on S^1.

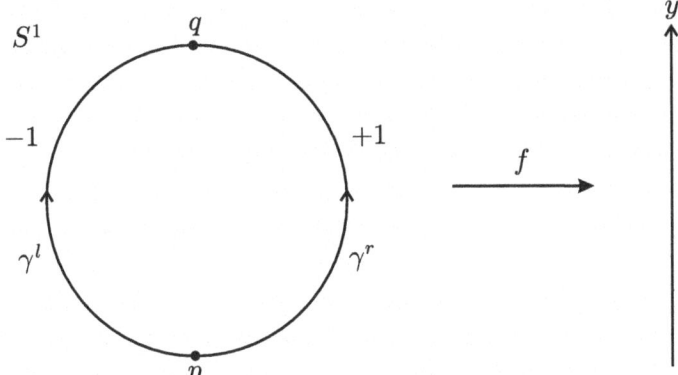

The height function on a circle

To see this directly, note that the isomorphism class of $\mathcal{L}$ is determined by the invertible linear map $\gamma_* : \mathbb{F} \to \mathbb{F}$ associated to the clockwise generator $[\gamma] \in \pi_1(S^1, q)$, i.e. an element $a \in \mathbb{F}^*$ where $\gamma_*(x) = ax$ and $\mathbb{F}$ is the ground field for the vector space. Choosing the constant path at q and the path from q to p on the right half of S^1 to define a local system associated to the representation, we have

$$\partial_1^{\mathcal{L}}(xq) = \left(\gamma_*^r(x) - \gamma_*^l(x)\right) p$$
$$= (x - ax) p.$$

Therefore, $H_k((C_*(f; \mathcal{L}), \partial_*^{\mathcal{L}})) \approx 0$ for all k whenever $a \neq 1$.

Example 6.20 (The Twisted Homology of a Torus) The torus $T^n = (S^1)^n$ is a topological group. Hence, for any Morse-Smale pair (f, g) and any local coefficient system $\mathcal{L}$ of rank one R-modules such that there exists a $[\gamma] \in \pi_1(T^n, e)$ with

$$\gamma_*(s) = gs \text{ for some } g \in R_e^\times \text{ where } g^{-1} - 1 \text{ is invertible},$$

$H_k((C_*(f; \mathcal{L}), \partial_*^{\mathcal{L}})) = 0$ for all $k = 0, \ldots, m$ by Theorem 6.16.

For the local coefficient system of rank one vector spaces e^η this can be seen directly using a computation similar to the one we present for the system of rank one Novikov modules on T^2 in Example 6.40. Note that this result is consistent with Example 6.10; although Example 6.10 was about twisted cohomology rather than twisted homology.

Example 6.21 (A Surface of Genus Two Is Not an Associative H-Space) In Example 6.12 we saw that the twisted Morse cohomology

$$H^*_{-\eta_1}(S) \approx H_k((C^*(f; e^{\eta_1}), \delta_*^{\eta_1}))$$

is nonzero, where e^{η_1} is a nontrivial local coefficient system of real line bundles on a surface S of genus 2. This suggests that the twisted Morse homology with coefficients in e^{η_1} should also be nonzero. However, the standard Universal Coefficient Theorems, which compare homology and cohomology do not apply to homology and cohomology with local coefficients, cf. Section V.7 of [17].

So, consider the η_1-twisted Morse-Smale-Witten chain complex

$$0 \xrightarrow{\partial_3^{\eta_1}} C_2(f; e^{\eta_1}) \xrightarrow{\partial_2^{\eta_1}} C_1(f; e^{\eta_1}) \xrightarrow{\partial_1^{\eta_1}} C_0(f; e^{\eta_1}) \longrightarrow 0$$

for the function and metric studied in Example 6.12. Using the notation from Example 6.12, $C_2(f; e^{\eta_1})$ consists of real multiples of p^2, $C_1(f; e^{\eta_1})$ consists of real linear combinations of p_1^1, p_2^1, p_3^1, and p_4^1, and $C_0(f; e^{\eta_1})$ consists of real multiples of p^0. Moreover, $\partial_2^{\eta_1}(p^2)$ is a nonzero multiple of p_2^1, the kernel of $\partial_1^{\eta_1}$ consists of real linear combinations of p_2^1, p_3^1, and p_4^1, and $\partial_1^{\eta_1}(p_1^1)$ is a nonzero multiple of p^0. Therefore,

$$H_k((C_*(f; e^{\eta_1}), \partial_*^{\eta_1})) \approx \begin{cases} 0 & \text{if } k = 0 \\ \mathbb{R} \oplus \mathbb{R} & \text{if } k = 1 \\ 0 & \text{if } k = 2, \end{cases}$$

and Corollary 6.17 implies that a surface of genus 2 is not an associative H-space. Note that this result also follows from Corollary 6.18.

Example 6.22 ($\mathbb{R}P^{2n}$ Is Not an Associative H-Space) Consider $M = \mathbb{R}P^2$ with the Morse-Smale function and orientations from Example 2.15. Since $H^1(\mathbb{R}P^2; \mathbb{R}) = 0$, the local coefficient system e^η is simple for every $\eta \in \Omega^1_{cl}(\mathbb{R}P^2, \mathbb{R})$. However, we can construct a non-simple local coefficient system $\mathcal{L}$ of rank one $\mathbb{R}$ vector spaces on $\mathbb{R}P^2$ using homomorphisms analogous to those defined in Example 2.15, where the nontrivial element $[\gamma] \in \pi_1(\mathbb{R}P^2, r)$ corresponds to the linear map $\gamma_* : \mathbb{R} \to \mathbb{R}$ given by $\gamma_*(x) = -x$.

$$(\gamma_{12,r})_*(x) = +x$$
$$(\gamma_{12,l})_*(x) = -x$$

6.2 H-Spaces

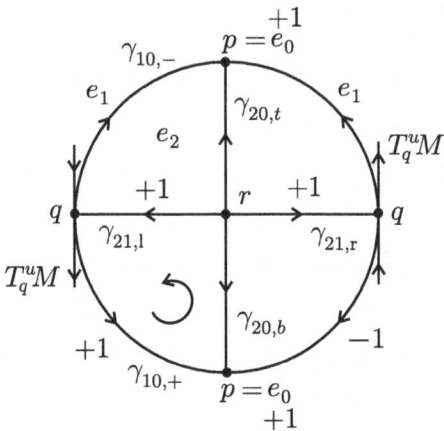

A Morse-Smale function on $\mathbb{R}P^2$

$$(\gamma_{01,+})_*(x) = +x$$
$$(\gamma_{01,-})_*(x) = -x$$

With this non-simple local system we have

$$H_2((C_*(f;\mathcal{L}), \partial_*^{\mathcal{L}})) \approx \mathbb{R}, \quad H_1((C_*(f;\mathcal{L}), \partial_*^{\mathcal{L}})) \approx 0, \quad H_0((C_*(f;\mathcal{L}), \partial_*^{\mathcal{L}})) \approx 0.$$

Therefore, Corollary 6.17 implies that $\mathbb{R}P^2$ is not an associative H-space.

A similar argument can be used to show that $\mathbb{R}P^n$ is not an associative H-space when n is even, but the obstruction gives no information when n is odd. That is, the function $\tilde{f} : S^n \to \mathbb{R}$ defined by

$$\tilde{f}(x_1, \ldots, x_n) = \sum_{j=2}^{n}(j-1)x_j^2$$

is a Morse-Smale function with respect to the standard Riemannian metric on S^n, and it satisfies $\tilde{f}(-x_1, \ldots, -x_n) = \tilde{f}(x_1, \ldots, x_n)$. So, $\tilde{f}$ induces a Morse-Smale function $f : \mathbb{R}P^n \to \mathbb{R}$, which has one critical point p_k of index k for all $k = 0, \ldots, n$. There are exactly two gradient flow lines between any two critical points of p_k and p_{k-1} of relative index one, and the signs associated to these two gradient flow lines will be the same when k is even and opposite when k is odd.

We can define a non-simple local coefficient system $\mathcal{L}$ of rank one $\mathbb{R}$ vector spaces on $\mathbb{R}P^n$ as above, where for all $k = 1, \ldots, n$ we have $(\gamma_1)_*(x) = x$ for one of the gradient flow lines from p_k to p_{k-1} and $(\gamma_2)_*(x) = -x$ for the other gradient flow line from p_k to p_{k-1}. Then the Morse-Smale-Witten chain complex with coefficients in $\mathcal{L}$ is

$$\mathbb{R} \xrightarrow{\partial_n} \mathbb{R} \xrightarrow{\partial_{n-1}} \cdots \xrightarrow{\partial_2} \mathbb{R} \xrightarrow{\partial_1} \mathbb{R} \longrightarrow 0,$$

where $\partial_k(x) = 0$ when k is even and $\partial_k(x) = 2x$ when k is odd. Therefore,

$$H_k((C_*(f;\mathcal{L}), \partial_*^{\mathcal{L}})) \approx 0$$

for all k when n is odd, but

$$H_n((C_*(f;\mathcal{L}), \partial_*^{\mathcal{L}})) \approx \mathbb{R}$$

when n is even. Thus, Corollary 6.17 implies that $\mathbb{R}P^{2n}$ is not an associative H-space.

Note The real projective space $\mathbb{R}P^n$ is an H-space only if $n = 1, 3,$ or 7, cf. Example 3C.3 of [36].

6.3 Novikov Homology

The construction of the (twisted or untwisted) Morse-Smale-Witten chain complex only required an exact 1-form $df \in \Omega^1(M, \mathbb{R})$, rather than a function $f : M \to \mathbb{R}$. That is, the critical points of f are the zeros of df, and the gradient flow lines of f are determined by df and a Riemannian metric on M. This observation naturally leads one to ask if it is possible to construct a chain complex analogous to the Morse-Smale-Witten chain complex using a closed Morse 1-form $\zeta \in \Omega^1_{cl}(M, \mathbb{R})$ in place of the exact Morse 1-form df, or at least obtain inequalities for ζ analogous to the Morse inequalities for df.

S.P. Novikov's influential paper [58] noted that a closed 1-form ζ on a differentiable manifold M defines a "multivalued function" S by integrating ζ over paths, and S becomes single valued on an appropriate covering space $\tilde{M} \to M$. The following problem is stated in [58]:

Problem To construct an analogue of Morse theory for the multivalued functions S. That is, to find a relationship between the stationary points $dS = 0$ of different index and the topology of the manifold M.

In the case where the integrals of the closed 1-form over cycles are integers, there is an associated $\mathbb{Z}$-covering and a circle valued function $f : M \to S^1$, cf. Lemma 2.1.17 of [61]. In this context, Novikov stated inequalities, now known as the Novikov inequalities, that generalize the Morse inequalities [58, 59]. A proof of the Novikov inequalities for general closed Morse 1-forms on finite dimensional closed smooth manifolds was given by M. Farber [28, 30].

The generalization of the Morse-Smale-Witten chain complex to closed 1-forms that determine integral cohomology classes, i.e. to circle valued Morse functions, was carried out by A. Pajitnov [64], and the construction of a Morse-Smale-Witten type complex using an arbitrary closed 1-form was given by D. Burghelea and S. Haller [18] and F. Latour [48]. These generalizations all define the boundary operator for the "Novikov complex" using the dynamics of a flow on a covering of the manifold determined by the closed 1-form and a Riemannian metric.

F. Farber and A. Ranicki used an alternate approach to construct an "algebraic Novikov complex" for a circle valued Morse functions using a noncommutative localization [31]. Their approach does not use the dynamics of a flow to define the boundary operator. Instead, they define their boundary operator using chain homotopy properties of the handle decomposition of certain cobordisms in a covering space of the manifold. Ranicki established relationships between the algebraic version of the Novikov complex and the dynamical version of the Novikov complex in certain cases [68, 69], and Farber extended the algebraic construction of the Novikov complex to general closed 1-forms [29, 30].

In this section we discuss how Novikov homology, i.e. the homology of the Novikov complexes, can be computed using a Morse complex with coefficients in a local system whose fiber is a Novikov ring.

6.3.1 A Covering Space Associated to a 1-Form

Novikov's idea for constructing a chain complex associated to a closed 1-form ζ on a smooth manifold M involves pulling back ζ to a covering space $\widetilde{M}_\xi$ of M where the pullback is exact. More specifically, if we assume M is connected and we pick a basepoint x_0 for M, then a closed 1-form $\zeta \in \Omega^1_{cl}(M, \mathbb{R})$ defines a homomorphism of periods $\mathrm{Per}_\xi : \pi_1(M, x_0) \to \mathbb{R}$ given by

$$\mathrm{Per}_\xi([\gamma]) = \int_\gamma \zeta,$$

which only depends on the de Rham cohomology class $\xi = [\zeta] \in H^1(M; \mathbb{R})$. The kernel $\Delta_\xi \stackrel{\mathrm{def}}{=} \ker \mathrm{Per}_\xi$ is a subgroup of $\pi_1(M, x_0)$, and hence it determines a covering space $\pi : \widetilde{M}_\xi \to M$ such that $\pi_\# : \pi_1(\widetilde{M}_\xi, \tilde{x}_0) \to \Delta_\xi \subseteq \pi_1(M, x_0)$ is an isomorphism, where $\tilde{x}_0 \in \widetilde{M}_\xi$ is any basepoint satisfying $\pi(\tilde{x}_0) = x_0$. In fact, $\pi_1(M, x_0)$ acts on the universal cover of M via deck transformations $\pi_1(M, x_0) \times \widetilde{M} \to \widetilde{M}$ and $\widetilde{M}_\xi = \widetilde{M}/\Delta_\xi$, cf. Theorem III.8.1 of [17]. Moreover, $\widetilde{M}_\xi$ is a regular covering space of M because $\ker \mathrm{Per}_\xi$ is a normal subgroup of $\pi_1(M, x_0)$, and the pullback $\tilde{\zeta} = \pi^*(\zeta)$ is exact on $\widetilde{M}_\xi$ because the function $\tilde{f} : \widetilde{M}_\xi \to \mathbb{R}$ given by

$$\tilde{f}(\tilde{x}) = \int_{\tilde{\gamma}} \tilde{\zeta},$$

where $\tilde{\gamma}$ is any path from $\tilde{x}_0$ to $\tilde{x}$, is well-defined and satisfies $d\tilde{f} = \tilde{\zeta}$.

Note If ζ is an **integer** valued 1-form, i.e. if $\xi = [\zeta] \in H^1(M; \mathbb{Z})$, then there is a well-defined circle valued function $f : M \to S^1$ given by

$$f(x) = e^{2\pi i \tilde{f}(\tilde{x})},$$

where $\tilde{x}$ is any element of the fiber $\pi^{-1}(x)$. Moreover, every smooth circle valued function $f : M \to S^1$ determines a corresponding smooth integer valued closed 1-form by pulling back $\frac{1}{2\pi} d\theta$ (see Example 2.14), where $[\frac{1}{2\pi} d\theta]$ is a generator for $H^1(S^1; \mathbb{Z})$. Thus, circle valued Morse theory can be viewed as a special case of Morse theory for closed 1-forms, cf. Lemma 2.1 of [30], Section III of [58], Section 6 of [59], or Section 2.1.4 of [61].

Definition 6.23 The **rank** of a cohomology class $\xi \in H^1(M; \mathbb{R})$ (or a closed 1-form) on a connected manifold M is defined to be the rank of the group of periods of ξ, i.e. the rank of the finitely generated abelian group $\Gamma_\xi \stackrel{\text{def}}{=} \text{Im Per}_\xi \subseteq \mathbb{R}$.

Since $H_1(M; \mathbb{Z})$ is isomorphic to $\pi_1(M, x_0)$ modulo its commutator subgroup and $\mathbb{R}$ is commutative, the period homomorphism Per_ξ factors through $H_1(M; \mathbb{Z})$. That is, there is a homomorphism $\overline{\text{Per}}_\xi : H_1(M; \mathbb{Z}) \to \mathbb{R}$ that makes the following diagram commute.

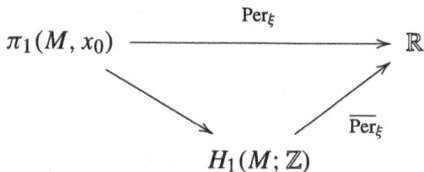

Hence, the rank of any cohomology class $\xi \in H^1(M; \mathbb{R})$ is bounded above by the rank of $H_1(M; \mathbb{Z})$, i.e. the number of elements in a basis for $H_1(M; \mathbb{Z})$ modulo its torsion subgroup.

Remark 6.24 (Rank One Cohomology Classes Are Dense in $H^1(M; \mathbb{R})$) If ξ is a **rank one** cohomology class, then we can divide ξ by the smallest positive element in Γ_ξ to get an integral cohomology class. Thus, rank one cohomology classes are real multiples of integral cohomology classes. Any rational cohomology class $\xi \in H^1(M; \mathbb{Q})$ is of rank one, because Im Per_ξ is generated by a finite set of rational numbers $\frac{a_1}{b_1}, \frac{a_2}{b_2}, \ldots, \frac{a_r}{b_r}$, which is a subgroup of the cyclic group generated by $\frac{1}{b_1 b_2 \cdots b_r}$. Thus, rank one cohomology classes are dense in $H^1(M; \mathbb{R})$, cf. Theorem 1.44 or Corollary 2.2 of [30]. Moreover, the Novikov homology of a general cohomology class $\xi = [\zeta] \in H^1(M; \mathbb{R})$ can be computed by perturbing a Morse form ζ to a rank one Morse form that agrees with ζ in a neighborhood of its critical points, cf. Section 4 of [77].

6.3 Novikov Homology

We now observe that the group of periods Γ_ξ is the quotient of $\pi_1(M, x_0)$ by the normal subgroup Δ_ξ, i.e.

$$\pi_1(M, x_0)/\Delta_\xi \approx \Gamma_\xi,$$

since Δ_ξ and Γ_ξ are the kernel and image of the homomorphism Per_ξ : $\pi_1(M, x_0) \to \mathbb{R}$. Moreover, $\pi_1(M, x_0)/\Delta_\xi$ is isomorphic to the deck transformation group of $\widetilde{M}_\xi$, cf. Corollary III.6.9 of [17] or Proposition 1.39 of [36], and hence the deck transformation group $\text{Aut}(\pi)$ of $\pi : \widetilde{M}_\xi \to M$ is isomorphic to Γ_ξ. The homomorphism $\text{Aut}(\pi) \to \Gamma_\xi$ can be described as follows. Starting with a deck transformation $D \in \text{Aut}(\pi)$, take any path $\tilde{\gamma}$ from the basepoint $\tilde{x}_0 \in \widetilde{M}_\xi$ to $D\tilde{x}_0$. The image of $\tilde{\gamma}$ under the projection $\pi : \widetilde{M}_\xi \to M$ maps to a loop γ in M based at x_0, and $\text{Per}_\xi([\gamma]) \in \Gamma_\xi$, which independent of the choice of the path $\tilde{\gamma}$.

This leads to the following, cf. Section 14.6 of [84].

Proposition 6.25 *The fiber and the deck transformation group of the regular covering space $\pi : \widetilde{M}_\xi \to M$ can be identified with the group of periods Γ_ξ.*

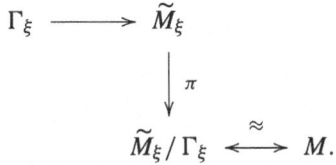

Proof Recall that $\widetilde{M}_\xi$ is a regular cover because $\pi_\#(\pi_1(\widetilde{M}_\xi, \tilde{x}_0)) = \Delta_\xi$ is the kernel of Per_ξ, and hence a normal subgroup of $\pi_1(M, x_0)$. Thus, the group of deck transformations of $\widetilde{M}_\xi$ acts transitively on the fibers of $\pi : \widetilde{M}_\xi \to M$. Moreover, a deck transformation is determined by where it maps the basepoint of $\widetilde{M}_\xi$, cf. Lemma III.4.4 of [17], and hence there is a bijection between the group of deck transformations and the fiber containing the basepoint. The fact that Γ_ξ is isomorphic to the deck transformation group of $\widetilde{M}_\xi$ completes the proof. □

Corollary 6.26 *If $\xi \in H^1(M; \mathbb{R})$ is a rank one cohomology class, then $\pi : \widetilde{M}_\xi \to M$ is a cyclic covering space.*

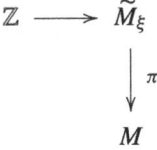

6.3.2 Novikov Rings

If M is compact, then every Morse form $\zeta \in \Omega^1_{cl}(M, \mathbb{R})$ will have finitely many zeros. However, the pullback $\tilde{\zeta}$ to $\tilde{M}_\xi$ of any closed Morse 1-form ζ with $0 \neq [\zeta] \in H^1(M; \mathbb{R})$ will have an infinite number of zeros, i.e. the Morse function $\tilde{f} : \tilde{M}_\xi \to \mathbb{R}$ will have infinitely many critical points, since the fiber Γ_ξ of $\tilde{M}_\xi$ is infinite. To account for this Novikov introduced a ring $\text{Nov}(\Gamma_\xi)$, where Γ_ξ is the group of periods of ζ.

Definition 6.27 Let $\Gamma \subseteq \mathbb{R}$ be an additive subgroup. The **Novikov ring** $\text{Nov}(\Gamma)$ is defined to be the ring of all formal power series of the form

$$\sum_{\gamma \in \Gamma} n_\gamma t^\gamma$$

with integer coefficients $n_\gamma \in \mathbb{Z}$, such that at most countably many $n_\gamma \neq 0$ and for any $c \in \mathbb{R}$ the set $\{\gamma \in \Gamma \mid n_\gamma \neq 0 \text{ and } \gamma > c\}$ is finite. The ring $\text{Nov}(\mathbb{R})$ will be denoted by Nov.

Note $\text{Nov}(\Gamma)$ is a commutative ring consisting of "half infinite" formal series with integer coefficients and exponents in $\Gamma \subseteq \mathbb{R}$. That is, every element $g \in \text{Nov}(\Gamma)$ can be written as $g = \sum_{i=1}^{\infty} n_i t^{\gamma_i}$, where $n_i \in \mathbb{Z}$, $\gamma_i \in \Gamma$, and $\gamma_1 > \gamma_2 > \gamma_3 > \cdots$. Addition and multiplication in $\text{Nov}(\Gamma)$ are defined formally, cf. Section 1.2.1 of [30], and an element of $\text{Nov}(\Gamma)$ is invertible if and only if its top coefficient is invertible in $\mathbb{Z}$, cf. Lemma 1.1.9 of [30] or Theorem 4.1 of [40]. Moreover, J.C. Sikorav proved that $\text{Nov}(\Gamma)$ is a principal ideal domain, cf. Lemma 1.1.10 of [30], Theorem 4.2 of [40], or Section 1 of [63].

Note If $\xi \in H^1(M; \mathbb{R})$ is a rank one cohomology class on a connected smooth manifold M, then $\Gamma_\xi \approx \mathbb{Z}$ and $\text{Nov}(\Gamma_\xi)$ is isomorphic to the Novikov ring $\text{Nov}(\mathbb{Z})$. That is, the ring with elements of the form

$$g = \sum_{i \in \mathbb{Z}} n_i t^i,$$

where only finitely many n_i with $i > 0$ are nonzero.

6.3.3 A Local Coefficient System of Rank One Nov-Modules

If M is connected and γ is a closed loop based at x_0, then $[\gamma] \in H_1(M; \mathbb{R})$ and

$$<\xi, [\gamma]> = \int_\gamma \zeta,$$

6.3 Novikov Homology

where $\xi = [\zeta] \in H^1(M; \mathbb{R})$. Moreover, $t^{<\xi,[\gamma]>} \in \text{Nov}(\Gamma_\xi)$ is invertible with inverse $t^{<\xi,[\gamma^{-1}]>} \in \text{Nov}(\Gamma_\xi)$. Thus, the cohomology class $\xi \in H^1(M; \mathbb{R})$ determines a representation

$$\pi_1(M, x_0) \times \text{Nov}(\Gamma_\xi) \to \text{Nov}(\Gamma_\xi)$$

defined by $(\gamma, g) \mapsto t^{<\xi,[\gamma]>} g$, and hence a bundle $\mathcal{L}(\Gamma_\xi)$ of rank one $\text{Nov}(\Gamma_\xi)$-modules on M (defined up to isomorphism).

If we extend the above representation to a representation on the Novikov ring $\text{Nov} \stackrel{\text{def}}{=} \text{Nov}(\mathbb{R})$, then we can give the following explicit description of a bundle $\mathcal{L}_\zeta$ in the isomorphism class $\mathcal{L}_\xi$ of bundles of rank one Nov-modules determined by $\xi = [\zeta] \in H^1(M; \mathbb{R})$.

Definition 6.28 For any $\zeta \in \Omega^1_{cl}(M, \mathbb{R})$ the local coefficient system $\mathcal{L}_\zeta$ of rank one Nov-modules on a connected smooth manifold M is defined as follows. The fiber at every point is Nov, and for every path $\gamma : [0, 1] \to M$, the isomorphism $\gamma_* : \text{Nov}_{\gamma(1)} \to \text{Nov}_{\gamma(0)}$ is defined to be multiplication by $t^{\int_0^1 \gamma^*(\zeta)}$, i.e.

$$g \mapsto t^{\int_0^1 \gamma^*(\zeta)} g.$$

Proposition 6.29 *Let $\zeta \in \Omega^1_{cl}(M, \mathbb{R})$, where M is a connected smooth manifold, and let $[\zeta] = \xi \in H^1(M; \mathbb{R})$. The local coefficient system $\mathcal{L}_\zeta$ is in the isomorphism class $\mathcal{L}_\xi$ of local coefficient systems determined by the representation*

$$\pi_1(M, x_0) \times \text{Nov} \to \text{Nov}$$

given by $(\gamma, g) \mapsto t^{<\xi,[\gamma]>} g$. In particular, if $\zeta_1, \zeta_2 \in \Omega^1_{cl}(M, \mathbb{R})$ represent the same de Rham cohomology class, then $\mathcal{L}_{\zeta_1}$ is isomorphic to $\mathcal{L}_{\zeta_2}$.

Proof The local coefficient system associated to the representation is defined by assigning Nov as the fiber at every point in M and fixing a homotopy class of paths rel endpoints from a basepoint $x_0 \in M$ to every point in M. Suppose that γ_0 represents the chosen homotopy class of paths from x_0 to $\gamma(0)$ and γ_1 represents the chosen homotopy class of paths from x_0 to $\gamma(1)$. The map $\gamma_* : \text{Nov}_{\gamma(1)} \to \text{Nov}_{\gamma(0)}$ is defined to be the homomorphism that the representation associates to the concatenated loop $\gamma_0 \gamma \gamma_1^{-1}$. (For more details see the proof of Theorem VI.1.12 in [93].) To complete the proof, note that the following diagram commutes.

$$\begin{array}{ccc} \text{Nov}_{\gamma(1)} & \xrightarrow{\times t^{\int_0^1 \gamma^*(\zeta)}} & \text{Nov}_{\gamma(0)} \\ {\scriptstyle \times t^{\int_0^1 \gamma_1^*(\zeta)}} \downarrow & & \downarrow {\scriptstyle \times t^{\int_0^1 \gamma_0^*(\zeta)}} \\ \text{Nov}_{\gamma(1)} & \xrightarrow{\times t^{<[\zeta],[\gamma_0 \gamma \gamma_1^{-1}]>}} & \text{Nov}_{\gamma(0)} \end{array}$$

□

Note The direction of integration for the 1-form in Definition 6.28 is opposite that for the 1-form η in the local system e^η defined in Example 2.5. The direction of integration for e^η was chosen so that η is integrated in the direction of the flow lines in Definition 2.12, whereas the direction of integration for the form ζ in Definition 6.28 was chosen to agree with that of $< [\zeta], [\gamma_0 \gamma \gamma_1^{-1}] >$.

6.3.4 Novikov Homology

Let ζ be a closed Morse 1-form on a closed connected smooth Riemannian manifold (M, g), and let Γ_ξ be its group of periods, where $\xi = [\zeta] \in H^1_{\text{dR}}(M; \mathbb{R})$. Since M is compact, the Morse form ζ has a finite number of zeros of Morse index k for all $k = 0, \ldots, m$. However, the flow associated to (ζ, g) doesn't necessarily behave like a gradient flow. For instance, there may be flow lines that begin and end at the same zero.

The lift of the form ζ and a generic metric g on M to the regular covering space $\pi : \widetilde{M}_\xi \to M$ associated to the kernel Δ_ξ of the period homomorphism determines a Morse-Smale function $\tilde{f} : \widetilde{M}_\xi \to \mathbb{R}$ whose critical points are the fibers above the zeros of ζ. In fact, the fiber above a zero of ζ of Morse index k consists of critical points of $\tilde{f}$ of Morse index k. Moreover, the fiber is in bijective correspondence with the group of periods Γ_ξ (Proposition 6.25). Thus, there are an infinite number of critical points of index k above every zero of index k whenever ζ is not exact, i.e. when the rank of ζ is greater than zero.

Various authors, including D. Burghelea and S. Haller [18], F. Latour [48], A. Pajitnov [64], M. Poźniak [65], and D. Schütz [75, 77], have defined Novikov complexes that are all similar to a Morse-Smale-Witten chain complex in the sense that the (co)chain groups in the complexes are free finitely generated modules over a ring and the boundary operator is defined by counting flow lines. However, the flow lines are counted on a regular covering space of M, such as $\widetilde{M}_\xi$, the universal cover $\widetilde{M}$, or a related cyclic cover $\bar{M}$ in the case where ζ has rank greater than one [77]. Moreover, the ring $\text{Nov}(\Gamma_\xi)$ is sometimes replaced with Nov [32, 76] or the Novikov ring associated to the related cyclic cover.

Let $\Gamma \subset \mathbb{R}$ be the group of periods for whatever covering space is being considered (Proposition 6.25). For each $k = 0, \ldots, m$, the degree k group in the Novikov complex $C_k(\zeta; \text{Nov}(\Gamma))$ is defined to be the free finitely generated $\text{Nov}(\Gamma)$-module on the index k zeros $Z_k(\zeta)$ of ζ, i.e.

$$C_k(\zeta; \text{Nov}(\Gamma)) \stackrel{\text{def}}{=} \left\{ \sum_{q \in Z_k(\zeta)} gq \;\middle|\; g \in \text{Nov}(\Gamma) \right\} \approx \bigoplus_{q \in Z_k(\zeta)} \text{Nov}(\Gamma),$$

and the boundary operator is defined intuitively using the Morse coefficients of the gradient flow of the Morse-Smale function $\tilde{f}$ defined by pulling ζ back to

6.3 Novikov Homology

the covering space (or of an $\tilde{f}$-gradient or a "gradient-like" flow associated to $\tilde{f}$). Alternately, $C_k(\zeta; \text{Nov})$ may be used in place of $C_k(\zeta; \text{Nov}(\Gamma))$

More specifically, above each zero q of ζ fix a critical point $\tilde{q}$ in the fiber of the covering space. The equivariant incidence coefficients are defined to be

$$<q:p> = \sum_{\gamma \in \Gamma} [\tilde{q} : \gamma \tilde{p}] t^{\gamma},$$

where $[\tilde{q} : \gamma \tilde{p}]$ denotes the number of flow lines from $\tilde{q}$ to $\gamma \tilde{p}$ counted with sign and we have identified $\Gamma \subseteq \mathbb{R}$ with the group of deck transformations of whatever cover is being used (see the discussion above Proposition 6.25). After proving that

$$<q:p> \in \text{Nov}(\Gamma) \subseteq \text{Nov},$$

$\partial_k^\zeta : C_k(\zeta; \text{Nov}(\Gamma)) \to C_{k-1}(\zeta; \text{Nov}(\Gamma))$ is defined on an elementary chain $gq \in C_k(\zeta; \text{Nov}(\Gamma_\xi))$ by

$$\partial_k^\zeta(gq) = \sum_{p \in Z_{k-1}(\zeta)} <q:p> gp,$$

cf. Section 1.4 of [18], Section 2 of [64], Section 2 of [65], or Section 4 of [77].

The following can be used to relate the homology of a Novikov chain complex to homology with local coefficients.

Novikov Principle [30, 32] *Given a closed 1-form ζ with Morse zeros on a finite dimensional closed connected smooth manifold M, let $\Gamma \subset \mathbb{R}$ be the group of periods of ζ. There exists a chain complex $(C_*(\zeta; \text{Nov}(\Gamma)), \partial_*^\zeta)$ of modules over the Novikov ring $\text{Nov}(\Gamma)$ with the following properties:*

1. *$C_k(\zeta; \text{Nov}(\Gamma))$ is a free finitely generated $\text{Nov}(\Gamma)$-module with a basis that is in one-to-one correspondence with the zeros of ζ of Morse index k for all $k = 0, \ldots, m$.*
2. *$(C_*(\zeta; \text{Nov}(\Gamma)), \partial_*^\zeta)$ is chain homotopy equivalent to a chain complex*

$$\text{Nov}(\Gamma) \otimes_{\mathbb{Z}[\Gamma]} C_*(\widetilde{M}_\xi),$$

where $C_*(\widetilde{M}_\xi)$ denotes the chain complex of cellular, simplicial, or singular, chains $\widetilde{M}_\xi$.

The Novikov Principle has been proved for several different types of Novikov complexes. A. Pajitnov proved the Novikov Principle for a Novikov complex defined using the flow defined by a closed Morse 1-form of rank one, i.e. a rational Morse form [64], and D. Schütz extended Pajitnov's approach to general closed Morse 1-forms using the fact that rank one cohomology classes are dense in $H^1(M; \mathbb{R})$ (Remark 6.24) [76, 77]. M. Farber and A. Ranicki also proved a version of the Novikov Principle for their "algebraic Novikov complex" for circle valued Morse functions [31], and Farber extended that approach to general closed Morse 1-forms [29, 30].

Remark 6.30 The Novikov Principle has been proved for rings more general than Nov(Γ) and Nov. The Novikov Principle has been proved for various completions of the group ring $\mathbb{Z}[\pi_1]$, where $\pi_1 = \pi_1(M, x_0)$, including the rational part of the Novikov ring, cf. Section 1.3 of [30], the noncommutative *Novikov-Sikorav completion* $\widehat{\mathbb{Z}[\pi_1]}_\xi$, cf. Section 3.1.5 of [30] and [79], and the noncommutative *universal Cohn localization* $\Sigma_\xi^{-1}(\mathbb{Z}[\pi_1])$, cf. [22] or Section 3.1.7 of [30]. The Cohn localization is used for the *Universal Novikov Principle*, which can be used to obtain the other applications of the Novikov Principle via extension of scalars. When using more general rings, the universal cover $\widetilde{M}$ is used in place of $\widetilde{M}_\xi$. That is, the Novikov Principle for a ring $\mathcal{R}$ says that there is a Novikov complex consisting of free $\mathcal{R}$-modules C_k^ζ, with a basis in bijective correspondence with the zeros of index k of a closed Morse 1-form ζ, that is chain homotopy equivalent to $\mathcal{R} \otimes_{\mathbb{Z}[\pi_1]} C_*(\widetilde{M})$.

Now suppose that $G \triangleleft \pi_1$ is a normal subgroup, where $\pi_1 = \pi_1(M, x_0)$. Let $\widetilde{M}_G \approx \widetilde{M}/G$ be the regular covering space of M corresponding to G, and let $\Gamma \approx \pi_1/G$ be the group of covering transformations of $\widetilde{M}_G$.

Proposition 6.31 *For any ring $\mathcal{R}$ and any ring homomorphism $\rho : \mathbb{Z}[\pi_1] \to \mathcal{R}$ that factors through the group ring $\mathbb{Z}[\Gamma]$ there is a chain equivalence*

$$\mathcal{R} \otimes_{\mathbb{Z}[\pi_1]} C_k(\widetilde{M}) \simeq \mathcal{R} \otimes_{\mathbb{Z}[\Gamma]} C_k(\widetilde{M}_G)$$

for all $k = 0, \ldots, m$, where $\pi_1 = \pi_1(M, x_0)$.

Proof Since $\widetilde{M}_G = \widetilde{M}/G$, an elementary singular chain $\sigma : \Delta^k \to \widetilde{M}$ determines an elementary singular chain $\pi \circ \sigma : \Delta^k \to \widetilde{M}_G$, where $\pi : \widetilde{M} \to \widetilde{M}_G$ is the projection map. Moreover, every elementary singular chain in $\widetilde{M}_G$ is of the form $\pi \circ \sigma$ for some elementary singular chain $\sigma : \Delta^k \to \widetilde{M}$, cf. Corollary III.4.2 of [17]. Thus, the projection map $\pi : \widetilde{M} \to \widetilde{M}_G$ induces a surjective chain map $\tilde{\pi} : \mathcal{R} \otimes_\mathbb{Z} C_k(\widetilde{M}) \to \mathcal{R} \otimes_\mathbb{Z} C_k(\widetilde{M}_G)$.

The assumption that ρ factors through $\mathbb{Z}[\Gamma]$, where $\Gamma \approx \pi_1/G$, means that there is a map $\bar{\rho}$ making the following diagram commute.

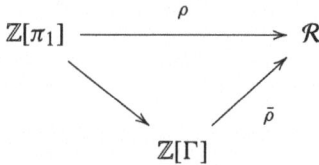

The chain group $\mathcal{R} \otimes_{\mathbb{Z}[\pi_1]} C_k(\widetilde{M})$ is the quotient of $\mathcal{R} \otimes_\mathbb{Z} C_k(\widetilde{M})$ by the subgroup generated by elements of the form

$$r\rho(\gamma) \otimes \sigma - r \otimes \gamma \cdot \sigma,$$

6.3 Novikov Homology

where $\gamma \in \pi_1$, $r \in \mathcal{R}$, and $\gamma \cdot \sigma$ denotes the action of the covering transformation determined by γ on the elementary singular chain $\sigma : \Delta^k \to \widetilde{M}$. Similarly, the chain group $\mathcal{R} \otimes_{\mathbb{Z}[\Gamma]} C_k(\widetilde{M}_G)$ is the quotient of $\mathcal{R} \otimes_{\mathbb{Z}} C_k(\widetilde{M}_G)$ by the subgroup generated by the elements of the form

$$r\bar{\rho}([\gamma]) \otimes \pi \circ \sigma - r \otimes [\gamma] \cdot (\pi \circ \sigma) = r\rho(\gamma) \otimes \pi \circ \sigma - r \otimes \pi \circ (\gamma \cdot \sigma),$$

where $[\gamma] \in \pi_1/G \approx \Gamma$, $r \in \mathcal{R}$, and $[\gamma] \cdot (\pi \circ \sigma)$ denotes the action of the covering transformation determined by $[\gamma]$ on the elementary singular chain $\pi \circ \sigma : \Delta^k \to \widetilde{M}_G$. Thus, there is a chain equivalence $\tilde{\pi} : \mathcal{R} \otimes_{\mathbb{Z}[\pi_1]} C_*(\widetilde{M}) \to \mathcal{R} \otimes_{\mathbb{Z}[\Gamma]} C_*(\widetilde{M}_G)$ making the following diagram commute for all $k = 0, \ldots, m$.

$$\begin{array}{ccc}
\mathcal{R} \otimes_{\mathbb{Z}} C_k(\widetilde{M}) & \xrightarrow{\tilde{\pi}} & \mathcal{R} \otimes_{\mathbb{Z}} C_k(\widetilde{M}_G) \\
\downarrow & & \downarrow \\
\mathcal{R} \otimes_{\mathbb{Z}[\pi_1]} C_k(\widetilde{M}) & \xrightarrow{\tilde{\pi}} & \mathcal{R} \otimes_{\mathbb{Z}[\Gamma]} C_k(\widetilde{M}_G)
\end{array}$$

□

The following corollary shows that the Novikov homology of a closed Morse 1-form $\zeta \in \Omega^1_{\text{cl}}(M, \mathbb{R})$ is isomorphic to the homology of M with local coefficients in the bundle $\mathcal{L}(\Gamma_\xi)$, where $[\zeta] = \xi \in H^1_{\text{dR}}(M; \mathbb{R})$.

Corollary 6.32 *If M is a finite dimensional closed connected smooth manifold and $\xi \in H^1_{\text{dR}}(M; \mathbb{R})$, then the following homology groups are isomorphic for all $k = 0, \ldots, m$.*

$$H_k((C_*(\zeta; \text{Nov}(\Gamma_\xi)), \partial^\zeta_*)) \approx H_k((\text{Nov}(\Gamma_\xi) \otimes_{\mathbb{Z}[\pi_1]} C_*(\widetilde{M}), \bar{\partial}_*)) \approx H_k(M; \mathcal{L}(\Gamma_\xi))$$

Proof By the Novikov Principle and Proposition 6.31 the following complexes are chain homotopy equivalent

$$(C_*(\zeta; \text{Nov}(\Gamma_\xi)), \partial^\zeta_*) \simeq (\text{Nov}(\Gamma_\xi) \otimes_{\mathbb{Z}[\Gamma_\xi]} C_*(\widetilde{M}_\xi), \bar{\partial}_*) \simeq (\text{Nov}(\Gamma_\xi) \otimes_{\mathbb{Z}[\pi_1]} C_*(\widetilde{M}), \bar{\partial}_*),$$

and by Eilenberg's Theorem (Theorem 4.3) there is an isomorphism between equivariant homology and homology with local coefficients, i.e.

$$H_k((\text{Nov}(\Gamma_\xi) \otimes_{\mathbb{Z}[\pi_1]} C_*(\widetilde{M}), \bar{\partial}_*)) \approx H_k(M; \mathcal{L}(\Gamma_\xi))$$

for all $k = 0, \ldots, m$. □

The preceding corollary and Theorem 4.1 show that Novikov homology can be computed using twisted Morse homology. However, the local coefficient system $\mathcal{L}(\Gamma_\xi)$ isn't well suited for use with twisted Morse homology, because it's only defined up to isomorphism and the exponents of the elements in $\text{Nov}(\Gamma_\xi)$ are limited to the elements in $\Gamma_\xi = \text{Im Per}_\xi$. For twisted Morse homology it's more convenient to use the local coefficient system $\mathcal{L}_\zeta$ from Definition 6.28, with fiber Nov instead of $\text{Nov}(\Gamma_\xi)$, because we can then integrate ζ over the gradient flow lines rather than loops created by arbitrarily picking homotopy classes of paths from a basepoint to the critical points of the Morse function, cf. Proposition 6.29.

The following theorem gives the relationship between $H_*(M; \mathcal{L}_\xi)$, where $\xi = [\zeta]$, and $H_*(M; \mathcal{L}(\Gamma_\xi))$.

Theorem 6.33 *If M is a finite dimensional closed connected smooth manifold and $\xi \in H^1_{dR}(M; \mathbb{R})$, then there is an isomorphism*

$$H_k(M; \mathcal{L}(\Gamma_\xi)) \otimes_{\text{Nov}(\Gamma_\xi)} \text{Nov} \approx H_k(M; \mathcal{L}_\xi)$$

for all $k = 0, \ldots, m$.

Proof Fix a basepoint $x_0 \in M$ and homotopy classes of paths rel endpoints from x_0 to every point in M. Also, pick some $\zeta \in \Omega^1_{cl}(M, \mathbb{R})$ such that $[\zeta] = \xi \in H^1_{dR}(M; \mathbb{R})$. These choices determine a local coefficient system $\mathcal{L}_\zeta(\Gamma_\xi)$ in the equivalence class $\mathcal{L}(\Gamma_\xi)$ and a local coefficient system $\mathcal{L}_\zeta(\mathbb{R})$ in the equivalence class $\mathcal{L}_\xi$, cf. Theorem VI.1.12 of [93]. The isomorphism $\gamma_* : \text{Nov}(\Gamma_\xi)_{\gamma(1)} \to \text{Nov}(\Gamma_\xi)_{\gamma(0)}$ associated to a path $\gamma : [0, 1] \to M$ by $\mathcal{L}_\zeta(\Gamma_\xi)$ is the restriction of the isomorphism $\gamma_* : \text{Nov}_{\gamma(1)} \to \text{Nov}_{\gamma(0)}$ associated to the path by $\mathcal{L}_\zeta(\mathbb{R})$. In both cases the isomorphism is given by multiplying by $t^{<\xi, [\gamma_0 \gamma \gamma_1^{-1}]>}$, where γ_0 and γ_1 are paths in the chosen homotopy classes of paths rel endpoints from the basepoint x_0 to $\gamma(0)$ and $\gamma(1)$ respectively.

Note that Nov is a $\text{Nov}(\Gamma_\xi)$-module because $\text{Nov}(\Gamma_\xi)$ is a subring of $\text{Nov}(\mathbb{R}) = \text{Nov}$. Thus,

$$\text{Nov}(\Gamma_\xi) \otimes_{\text{Nov}(\Gamma_\xi)} \text{Nov} \approx \text{Nov},$$

cf. Theorem 5.7 of [42] or Proposition 2.58 of [72]. So, picking any Morse-Smale pair (f, g) on M we have

$$C_k(f; \mathcal{L}_\zeta(\Gamma_\xi)) \otimes_{\text{Nov}(\Gamma_\xi)} \text{Nov} \approx C_k(f; \mathcal{L}_\zeta(\mathbb{R})),$$

and hence,

$$H_k((C_*(f; \mathcal{L}_\zeta(\Gamma_\xi)) \otimes_{\text{Nov}(\Gamma_\xi)} \text{Nov}, \partial_*)) \approx H_k((C_*(f; \mathcal{L}_\zeta(\mathbb{R})), \partial_*))$$

for all $k = 0, \ldots, m$, since $\partial_*^{\mathcal{L}_\zeta(\Gamma_\xi)}$ is simply a restriction of $\partial_*^{\mathcal{L}_\zeta(\mathbb{R})}$.

Now recall that $\text{Nov}(\Gamma_\xi)$ is a principle ideal domain that is torsion free, cf. Lemma 1.10 and Lemma 1.12 of [30]. Thus, $(C_*(f; \mathcal{L}_\zeta(\Gamma_\xi)), \partial_*)$ is a complex of flat $\text{Nov}(\Gamma_\xi)$-modules whose subcomplex of boundaries is also flat, cf. Corollary

3.50 of [72]. Therefore, the Universal Coefficient Theorem, cf. Theorem 7.55 of [72], gives a short exact sequence

$$0 \longrightarrow H_k((C_*(f; \mathcal{L}_\zeta(\Gamma_\xi)), \partial_*)) \otimes_{\mathrm{Nov}(\Gamma_\xi)} \mathrm{Nov} \xrightarrow{\lambda_k} H_k((C_*(f; \mathcal{L}_\zeta(\mathbb{R})), \partial_*))$$

$$\longrightarrow \mathrm{Tor}_1^{\mathrm{Nov}(\Gamma_\xi)}(H_{k-1}((C_*(f; \mathcal{L}_\zeta(\Gamma_\xi)), \partial_*)), \mathrm{Nov}) \longrightarrow 0$$

where $\lambda_k([c] \otimes g) = [c \otimes g]$, and the $\mathrm{Tor}_1^{\mathrm{Nov}(\Gamma_\xi)}$ term is zero because Nov is flat as a Nov(Γ_ξ)-module, cf. Lemma 1.12 of [30] and Theorem 7.2 of [72]. Therefore,

$$H_k((C_*(f; \mathcal{L}_\zeta(\Gamma_\xi)), \partial_*)) \otimes_{\mathrm{Nov}(\Gamma_\xi)} \mathrm{Nov} \approx H_k((C_*(f; \mathcal{L}_\zeta(\mathbb{R})), \partial_*)),$$

and by the Morse Homology Theorem (Theorem 4.1) we have

$$H_k(M; \mathcal{L}(\Gamma_\xi)) \otimes_{\mathrm{Nov}(\Gamma_\xi)} \mathrm{Nov} \approx H_k(M; \mathcal{L}_\xi)$$

for all $k = 0, \ldots, m$. $\square$

6.3.5 Novikov Numbers

The following lemma allows us to define the Novikov numbers associated to a cohomology class $\xi \in H^1(M; \mathbb{R})$, cf. Section 1.5 of [30].

Lemma 6.34 *Let M be a closed connected finite dimensional smooth manifold of dimension m, and let $\xi \in H^1(M; \mathbb{R})$. For all $k = 0, \ldots, m$, $H_k(M; \mathcal{L}_\xi)$ is a finitely generated Nov-module, and hence it is isomorphic to a finitely generated free Nov-module and finitely many cyclic torsion modules.*

Proof Picking any $\zeta \in \Omega^1_{\mathrm{cl}}(M, \mathbb{R})$ such that $[\zeta] = \xi \in H^1(M; \mathbb{R})$ gives a bundle of rank one Nov-modules $\mathcal{L}_\zeta$ in the isomorphism class $\mathcal{L}_\xi$. For any Morse-Smale pair (f, g) the chain groups in the twisted Morse-Smale-Witten chain complex $(C_*(f; \mathcal{L}_\zeta), \partial_*^{\mathcal{L}_\zeta})$ are finitely generated Nov-modules, and

$$H_k((C_*(f; \mathcal{L}_\zeta), \partial_*^{\mathcal{L}_\zeta})) \approx H_k(M; \mathcal{L}_\xi)$$

for all $k = 0, \ldots, m$ by the Twisted Morse Homology Theorem (Theorem 4.1), whose proof also holds in the category of modules over a ring. Therefore, $H_k(M; \mathcal{L}_\xi)$ is a finitely generated Nov-module. The second claim follows from the structure theorem for finitely generated modules over principal ideal domains, cf. Theorem 6.12 of [42] or Corollary 1.5.3 of [74]. $\square$

Definition 6.35 Let $\xi \in H^1(M;\mathbb{R})$. The **Novikov numbers** $b_k(\xi)$ and $q_k(\xi)$ are defined as follows for all $k = 0,\ldots, m$.

$b_k(\xi) =$ the rank of $H_k(M;\mathcal{L}_\xi)$ as a module over Nov.

$q_k(\xi) =$ the minimal number of generators of the torsion

submodule of $H_k(M;\mathcal{L}_\xi)$.

The Novikov numbers generalize the Betti numbers and torsion numbers of a manifold.

Proposition 6.36 (Proposition 1.28 [30]) *If $\xi = 0 \in H^1(M;\mathbb{R})$, then the Novikov number $b_k(\xi)$ coincides with the Betti number $b_k(M)$, i.e. the rank of $H_k(M;\mathbb{Z})$, and $q_k(\xi)$ coincides with the minimal number of generators of the torsion subgroup of $H_k(M;\mathbb{Z})$, for all $k = 0,\ldots,m$.*

Proof The local coefficient system $\mathcal{L}_0$ is constant, since the homomorphisms are given by multiplication by t^0, and hence $H_k(M;\mathcal{L}_0) = H_k(M;\text{Nov})$ for all $k = 0,\ldots,m$. By the Universal Coefficient Theorem we have the following split short exact sequence of abelian groups, cf. Corollary 3A.4 of [36], Theorem V.2.5 of [38], or Theorem 7.55 of [72]

$$0 \longrightarrow H_k(M;\mathbb{Z}) \otimes \text{Nov} \xrightarrow{\lambda_k} H_k(M;\text{Nov}) \longrightarrow \text{Tor}(H_{k-1}(M;\mathbb{Z}),\text{Nov}) \longrightarrow 0$$

where the Tor term is zero since Nov is a torsion free abelian group, cf. Proposition 3A.5 of [36] or Proposition 3.1.4 of [92]. Moreover, λ_k is defined by $\lambda_k([c] \otimes g) = [c \otimes g]$, which is a homomorphism of Nov-modules. Hence λ_k gives an isomorphism of Nov-modules

$$H_k(M;\text{Nov}) \approx H_k(M;\mathbb{Z}) \otimes \text{Nov}$$

for all $k = 0,\ldots,m$. □

6.3.6 Novikov Inequalities

A closed 1-form $\zeta \in \Omega^1_{\text{cl}}(M,\mathbb{R})$ is locally exact. In particular, for every zero $p \in M$ of ζ there exists a neighborhood U of p such that $\zeta|_U = df$ for some function $f : U \to \mathbb{R}$. A zero p of ζ is a critical point of f, and hence the notions of nondegenerate and index for critical points of a function carry over to the zeros of a closed 1-form. That is, a zero p of ζ is called **nondegenerate** if and only if p is a nondegenerate critical point of f, and the **Morse index** of the zero p is defined to be the Morse index of p as a critical point of f.

The significance of the Novikov numbers is indicated by the following theorem, which generalizes the Morse inequalities. The theorem was proved by M. Farber in the case where ζ has rank one in [28]. The theorem can then be proved for a closed 1-form ζ of higher rank by perturbing ζ to a closed form of rank one having the same zeros as ζ, cf. Lemma 2.5 of [30]. The theorem also follows from the more general Novikov Principle.

Theorem 6.37 (Novikov Inequalities) *Let $\zeta \in \Omega^1_{cl}(M, \mathbb{R})$, and assume that all the zeros of ζ are nondegenerate. If $c_k(\zeta)$ denotes the number of zeros of ζ with Morse index k, then*

$$c_k(\zeta) \geq b_k(\xi) + q_k(\xi) + q_{k-1}(\xi)$$

for all $k = 0, \ldots, m$, where $\xi = [\zeta] \in H^1_{dR}(M; \mathbb{R})$.

6.3.7 Novikov Numbers and Twisted Morse Homology

The Twisted Morse Homology Theorem (Theorem 4.1) and Proposition 6.29 and imply that the Novikov numbers can be computed using a Morse-Smale-Witten chain complex with coefficients in the bundle of rank one Nov-modules $\mathcal{L}_\zeta$ defined in Definition 6.28.

Proposition 6.38 *Let $f : M \to \mathbb{R}$ be a smooth Morse-Smale function on a closed connected finite dimensional smooth Riemannian manifold (M, g). If $\xi \in H^1(M; \mathbb{R})$ and $\zeta \in \Omega^1_{cl}(M, \mathbb{R})$ satisfies $[\zeta] = \xi$, then*

$b_k(\xi) =$ *the rank of $H_k((C_*(f; \mathcal{L}_\zeta), \partial_*^{\mathcal{L}_\zeta}))$ as a module over Nov,*

$q_k(\xi) =$ *the minimal number of generators of the torsion*

submodule of $H_k((C_*(f; \mathcal{L}_\zeta), \partial_*^{\mathcal{L}_\zeta}))$,

for all $k = 0, \ldots, m$.

We now present a few concrete examples where we compute the Novikov numbers using twisted Morse homology.

Example 6.39 (The Novikov Numbers of a Circle) Consider a closed 1-form ζ on the unit circle $S^1 \subset \mathbb{R}^2$. Using the Morse-Smale function and orientations from Example 2.14 we can compute the Novikov numbers $b_k(\xi)$ and $q_k(\xi)$ using the following twisted Morse-Smale-Witten complex with coefficients in $\mathcal{L}_\zeta$, where $\xi = [\zeta] \in H^1(S^1; \mathbb{R})$.

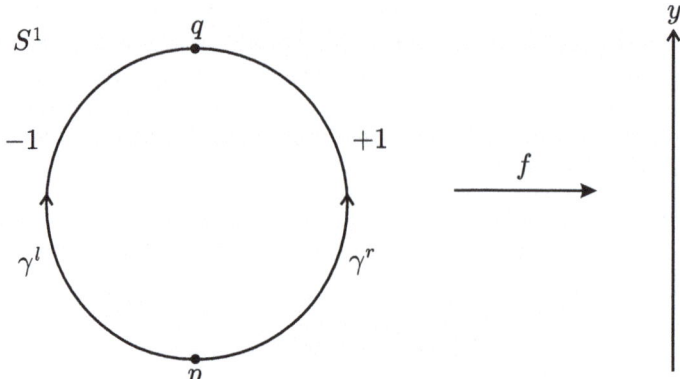

The height function on a circle

$$0 \longrightarrow C_1(f;\mathcal{L}_\zeta) \xrightarrow{\partial_1^{\mathcal{L}_\zeta}} C_0(f;\mathcal{L}_\zeta) \longrightarrow 0$$

$$\downarrow \approx \qquad\qquad \downarrow \approx$$

$$0 \longrightarrow \mathrm{Nov}_q \xrightarrow{\partial_1^{\mathcal{L}_\zeta}} \mathrm{Nov}_p \longrightarrow 0$$

We have

$$\partial_1^{\mathcal{L}_\zeta}(gq) = \left(t^{\int_0^1 (\gamma^r)^*(\zeta)} g - t^{\int_0^1 (\gamma^l)^*(\zeta)} g\right) p$$

for all $g \in \mathrm{Nov}$, and hence if ζ is exact

$$H_k((C_*(f;\mathcal{L}_\zeta), \partial_*^{\mathcal{L}_\zeta})) \approx \begin{cases} \mathrm{Nov} & \text{if } k = 0, 1 \\ 0 & \text{otherwise.} \end{cases}$$

Therefore, $H_k((C_*(f;\mathcal{L}_\zeta), \partial_*^{\mathcal{L}_\zeta})) \approx H_k(S^1; \mathbb{Z}) \otimes \mathrm{Nov}$ for all k when $[\zeta] = \xi = 0$, since $\mathbb{Z} \otimes \mathrm{Nov} \approx \mathrm{Nov}$, cf. Theorem 5.7 of [42]. Thus, $b_k(0)$ and $q_k(0)$ agree with the Betti numbers and torsion numbers of S^1 for all k, cf. Proposition 6.36.

Now consider the case where $\zeta = d\theta$ is the non exact closed 1-form from Example 2.14. For any $g \in \mathrm{Nov}$ we have

$$\partial_1^{\mathcal{L}_\zeta}(gq) = \left(t^{\int_0^1 (\gamma^r)^*(d\theta)} g - t^{\int_0^1 (\gamma^l)^*(d\theta)} g\right) p = \left((t^\pi - t^{-\pi}) g\right) p,$$

where $t^\pi - t^{-\pi}$ is invertible since

$$\frac{1}{t^\pi - t^{-\pi}} = \frac{1}{t^\pi} \frac{1}{1 - t^{-2\pi}} = \sum_{i=0}^{\infty} t^{-2\pi i - \pi} \in \mathrm{Nov}.$$

6.3 Novikov Homology

Hence, $\partial_1^{\mathcal{L}_\zeta} : C_1(f; \mathcal{L}_\zeta) \to C_0(f; \mathcal{L}_\zeta)$ is surjective, and $H_k((C_*(f; \mathcal{L}_\zeta), \partial_*^{\mathcal{L}_\zeta})) \approx 0$ for all k. Thus, $b_k([d\theta]) = q_k([d\theta]) = 0$ for all k. Similarly, one can show that $b_k(\xi) = q_k(\xi) = 0$ for all k whenever $\xi \neq 0$.

Note If we change the orientation of $W^u(q)$ in the previous example to be counterclockwise instead of clockwise, then for all $g \in \text{Nov}$

$$\partial_1^{\mathcal{L}_\zeta}(gq) = \left(t^{\int_0^1 (\gamma^l)^*(d\theta)} g - t^{\int_0^1 (\gamma^r)^*(d\theta)} g\right) p = \left((t^{-\pi} - t^\pi) g\right) p,$$

and the element $t^{-\pi} - t^\pi \in \text{Nov}$ is still invertible because

$$\frac{1}{t^{-\pi} - t^\pi} = \frac{-1}{t^\pi} \frac{1}{1 - t^{-2\pi}} = \sum_{i=0}^\infty (-1) t^{-2\pi i - \pi} \in \text{Nov}.$$

In fact, as noted earlier, an element in Nov is invertible if and only if its top coefficient is ± 1, cf. Lemma 1.9 of [30].

Example 6.40 (The Novikov Numbers of a Torus) Consider a closed 1-form ζ on the torus $T^2 = S^1 \times S^1$, viewed as a square with opposite edges identified. Using the construction detailed in the proof of Theorem 4.12 we can create a Morse-Smale pair (f, g) on T^2 with one critical point p of index 0, two critical points q, r of index 1, and one critical point s of index 2, such that there are exactly two gradient flow lines between each pair of critical points of relative index one. If we orient each S^1 clockwise, give $S^1 \times S^1$ the product orientation, and follow the orientation conventions in Sect. 2.2 for the moduli spaces, then the signs associated to the gradient flow lines are as indicated in the diagram.

For a pair of critical points (x_1, x_2) with $\lambda_{x_2} - \lambda_{x_1} = 1$, let $\gamma^+_{x_1 x_2}$ and $\gamma^-_{x_1 x_2}$ denote parameterizations of the gradient flow lines from x_1 to x_2 with the signs $+1$ and -1 respectively. The twisted Morse-Smale-Witten chain complex $(C_*(f; \mathcal{L}_\zeta), \partial_*^{\mathcal{L}_\zeta})$ is

$$\begin{array}{ccccccccc}
0 & \longrightarrow & C_2(f; \mathcal{L}_\zeta) & \xrightarrow{\partial_2^{\mathcal{L}_\zeta}} & C_1(f; \mathcal{L}_\zeta) & \xrightarrow{\partial_1^{\mathcal{L}_\zeta}} & C_0(f; \mathcal{L}_\zeta) & \longrightarrow & 0 \\
 & & \updownarrow \approx & & \updownarrow \approx & & \updownarrow \approx & & \\
0 & \longrightarrow & \text{Nov}_s & \xrightarrow{\partial_2^{\mathcal{L}_\zeta}} & \text{Nov}_r \oplus \text{Nov}_q & \xrightarrow{\partial_1^{\mathcal{L}_\zeta}} & \text{Nov}_p & \longrightarrow & 0
\end{array}$$

where the boundary operator is given by

$$\partial_k^{\mathcal{L}_\zeta}(gx_2) = \sum_{x_1 \in Cr_{k-1}(f)} \left(t^{\int_0^1 (\gamma^+_{x_1 x_2})^*(\zeta)} g - t^{\int_0^1 (\gamma^-_{x_1 x_2})^*(\zeta)} g\right) x_1,$$

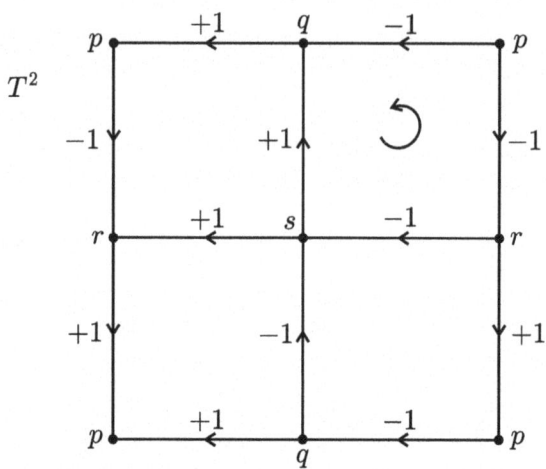

Orientations for unstable manifolds on $T^2 = S^1 \times S^1$

for $k = \lambda_{x_2}$ and $g \in \text{Nov}$. If ζ is exact, then $\partial_k^{\mathcal{L}_\zeta} = 0$ and

$$H_k((C_*(f;\mathcal{L}_\zeta), \partial_*^{\mathcal{L}_\zeta})) \approx H_k(T^2; \mathbb{Z}) \otimes \text{Nov}$$

for all k, since $\mathbb{Z} \otimes \text{Nov} \approx \text{Nov}$, cf. Theorem 5.7 of [42]. Hence, $b_k(0)$ and $q_k(0)$ agree with the Betti numbers and torsion numbers of T^2 for all k, cf. Proposition 6.36.

Now consider the case where $\zeta = \pi_1^*(d\theta)$ is the pullback of $d\theta$, the non exact closed 1-form from Example 2.14, under the projection onto the first factor of $T^2 = S^1 \times S^1 \subset \mathbb{R}^2 \times \mathbb{R}^2$. In this case we have

$$\partial_1^{\mathcal{L}_\zeta}(gq) = \left(t^{\int_0^1 (\gamma_{pr}^+)^*(\zeta)} g - t^{\int_0^1 (\gamma_{pr}^-)^*(\zeta)} g \right) p = ((t^\pi - t^{-\pi})g)p,$$

and

$$\partial_2^{\mathcal{L}_\zeta}(gs) = \left(t^{\int_0^1 (\gamma_{qs}^+)^*(\zeta)} g - t^{\int_0^1 (\gamma_{qs}^-)^*(\zeta)} g \right) q = ((t^\pi - t^{-\pi})g)r,$$

and $\partial_1^{\mathcal{L}_\zeta}(gr) = 0$ for all $g \in \text{Nov}$. Since $t^\pi - t^{-\pi}$ is invertible in Nov, this implies that

$$H_k((C_*(f;\mathcal{L}_\zeta), \partial_*^{\mathcal{L}_\zeta})) \approx 0$$

for all k, and hence, $b_k([\zeta]) = q_k([\zeta]) = 0$ for all k when $\zeta = \pi_1^*(d\theta)$. This is also true when $\zeta = \pi_2^*(d\theta)$, and in fact, for any $\xi \in H^1(T^2; \mathbb{R})$ with $\xi \neq 0$. This can be seen directly using a computation similar to the one presented above, or by using the fact that T^2 is an associative H-space and $t^\gamma - 1$ is invertible in Nov for all $\gamma \in \mathbb{R}$ with $\gamma \neq 0$, cf. Example 6.20.

6.3 Novikov Homology

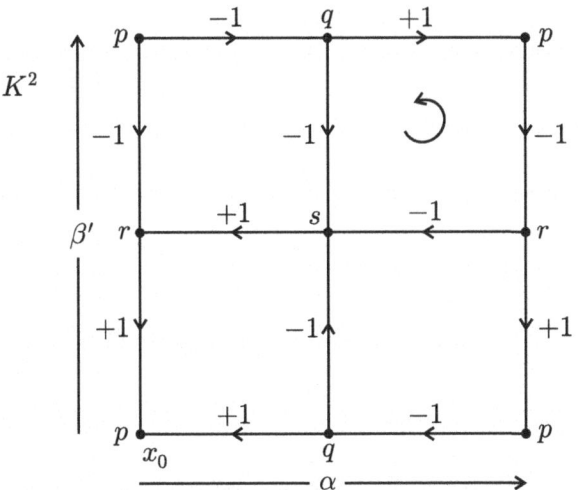

Orientations for unstable manifolds on a Klein bottle

Example 6.41 (The Novikov Numbers of a Klein Bottle) Consider the Klein bottle K^2 as $\mathbb{R}^2$ modulo the properly discontinuous action of the group Δ generated by

$$\alpha(x, y) = (x + 1, y)$$
$$\beta(x, y) = (1 - x, y + 1).$$

Then $K^2 = \mathbb{R}^2/\Delta$ is a compact smooth manifold with universal cover $\mathbb{R}^2$ and deck transformation group Δ. Moreover, the fundamental group of K^2 is

$$\pi_1(K^2, x_0) \approx \Delta \approx \{a, b | \, b^{-1}ab = a^{-1}\},$$

cf. Corollary III.7.3 and Example III.7.5 of [17]. Since $H_1(K^2; \mathbb{Z})$ is the abelianization of $\pi_1(K^2, x_0)$, this implies that

$$H_1(K^2; \mathbb{Z}) \approx \{a, b | a^2 = 1\} \approx \mathbb{Z} \oplus \mathbb{Z}_2,$$

and hence $H_1(K^2; \mathbb{R}) \approx \mathbb{R}$. If we consider $U = [0, 1] \times [0, 1] \subset \mathbb{R}^2$ as a fundamental domain for the action of Δ with basepoint $x_0 = (0, 0)$, then $\alpha \in \pi_1(K^2, x_0)$ corresponds to a path along the bottom edge of U and $\beta \in \pi_1(K^2, x_0)$ corresponds to a diagonal in U path from $(0, 0)$ to $(1, 1)$, cf. Definition III.6.7 of [17]. The diagonal path from $(0, 0)$ to $(1, 1)$ is homotopic rel endpoints to a concatenated path along the boundary of U from $(0, 0)$ to $(0, 1)$ to $(1, 1)$, and since homology with real coefficients has no torsion, a path β' along the left edge of U from $(0, 0)$ to $(0, 1)$ represents a generator of $H_1(K^2; \mathbb{R}) \approx \mathbb{R}$.

Using the construction detailed in the proof of Theorem 4.12 we can create a Morse-Smale pair $(f, \mathfrak{g})$ on K^2 with one critical point s of index 2, two critical points q, r of index 1, and one critical point p of index 0, such that there are exactly two gradient flow lines between each pair of critical points of relative index one. If $\zeta \in \Omega^1_{\mathrm{cl}}(K, \mathbb{R})$ is exact, then using the orientations shown in the diagram we have $\partial^{\mathcal{L}_\zeta}_1 = 0$ and for any $g \in \mathrm{Nov}$

$$\partial^{\mathcal{L}_\zeta}_2(gs) = (t^{\gamma_r}g - t^{\gamma_r}g)r + (-t^{\gamma_q}g - t^{\gamma_q}g)q = -2t^{\gamma_q}gq,$$

for some $\gamma_r, \gamma_q \in \mathbb{R}$, where $-2t^{\gamma_q}$ is not invertible in Nov. This implies that

$$H_k((C_*(f; \mathcal{L}_\zeta), \partial^{\mathcal{L}_\zeta}_*)) \approx \begin{cases} \mathrm{Nov} & \text{if } k = 0 \\ \mathrm{Nov} \oplus \mathrm{Nov}/2\mathrm{Nov} & \text{if } k = 1 \\ 0 & \text{otherwise.} \end{cases}$$

Hence, $b_0(0) = b_1(0) = q_1(0) = 1$, and $b_k(0) = q_k(0) = 0$ for all other k, cf. Proposition 6.36.

Now let ζ be a closed 1-form such that $\int_{\beta'} \zeta = 2\pi$ and $\int_\alpha \zeta = 0$. The form ζ exists because β' represents the generator of $H_1(K^2; \mathbb{R})$. Explicitly, we can choose ζ to be the pullback of the closed 1-form $d\theta$ on S^1 from Example 2.14 along the projection onto the y-axis in the above diagram. With this choice of $\zeta \in \Omega_{\mathrm{cl}}(K^2, \mathbb{R})$ we have

$$\partial^{\mathcal{L}_\zeta}_1(gr) = ((t^\pi - t^{-\pi})g)p \quad \text{and} \quad \partial^{\mathcal{L}_\zeta}_2(gs) = ((-t^\pi - t^{-\pi})g)q,$$

for any $g \in \mathrm{Nov}$, where both $t^\pi - t^{-\pi}$ and $-t^\pi - t^{-\pi}$ are invertible in Nov because their leading coefficients are invertible in $\mathbb{Z}$. Therefore, $H_k((C_*(f; \mathcal{L}_\zeta), \partial^{\mathcal{L}_\zeta}_*)) \approx 0$ for all k, and hence $b_k([\zeta]) = q_k([\zeta]) = 0$ for all k. The same is true when ζ is any non exact closed 1-form. Note that the fundamental difference between the computations when ζ is exact and when ζ is not exact is that $-2t^\gamma$ is not invertible in Nov for any $\gamma \in \mathbb{R}$, but $-t^\gamma - t^{\gamma-a}$ is invertible in Nov for all $\gamma, a \in \mathbb{R}$ with $a \neq 0$.

Example 6.42 (The Novikov Numbers of a Surface of Genus Two) Let S be the surface of genus 2 from Example 6.12, with the same Morse-Smale pair $(f, \mathbf{g})$ and basis of closed 1-forms $\{\eta_1, \eta_2, \eta_3, \eta_4\}$ for $H^1_{\mathrm{dR}}(S; \mathbb{R})$.

For a pair of critical points (x_1, x_2) with $\lambda_{x_2} - \lambda_{x_1} = 1$, let $\gamma^+_{x_1 x_2}$ and $\gamma^-_{x_1 x_2}$ denote parameterizations of the gradient flow lines from x_1 to x_2 with the signs $+1$ and -1 respectively. Let ζ be a closed 1-form on S and consider the twisted Morse-Smale-Witten chain complex with coefficients in $\mathcal{L}_\zeta$,

$$\begin{array}{ccccccccc}
0 & \longrightarrow & C_2(f; \mathcal{L}_\zeta) & \xrightarrow{\partial^{\mathcal{L}_\zeta}_2} & C_1(f; \mathcal{L}_\zeta) & \xrightarrow{\partial^{\mathcal{L}_\zeta}_1} & C_0(f; \mathcal{L}_\zeta) & \longrightarrow & 0 \\
& & \updownarrow \approx & & \updownarrow \approx & & \updownarrow \approx & & \\
0 & \longrightarrow & \mathrm{Nov}_{p^2} & \xrightarrow{\partial^{\mathcal{L}_\zeta}_2} & \mathrm{Nov}_{p^1_1} \oplus \mathrm{Nov}_{p^1_2} \oplus \mathrm{Nov}_{p^1_3} \oplus \mathrm{Nov}_{p^1_4} & \xrightarrow{\partial^{\mathcal{L}_\zeta}_1} & \mathrm{Nov}_{p^0} & \longrightarrow & 0
\end{array}$$

6.3 Novikov Homology

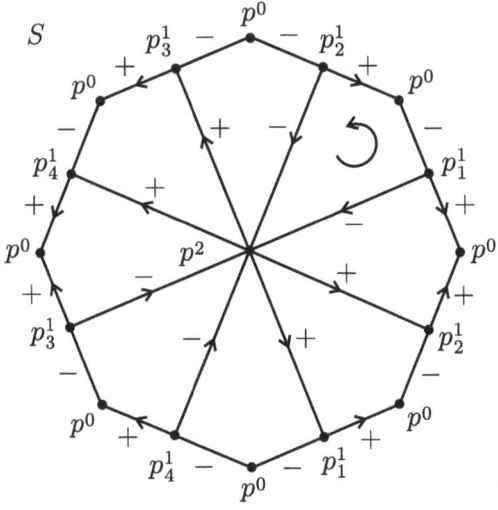

A surface of genus 2

where the boundary operator is given by

$$\partial_k^{\mathcal{L}_\zeta}(gx_2) = \sum_{x_1 \in Cr_{k-1}(f)} \left(t^{\int_0^1 (\gamma_{x_1 x_2}^+)^*(\zeta)} g - t^{\int_0^1 (\gamma_{x_1 x_2}^-)^*(\zeta)} g \right) x_1$$

for $k = \lambda_{x_2}$ and $g \in \text{Nov}$. If ζ is exact, then $\partial_k^{\mathcal{L}_\zeta} = 0$ and

$$H_k((C_*(f; \mathcal{L}_\zeta), \partial_*^{\mathcal{L}_\zeta})) \approx H_k(S; \mathbb{Z}) \otimes \text{Nov}$$

for all k. Thus, $b_0(0) = 1$, $b_1(0) = 4$, $b_2(0) = 1$, $b_k(0) = 0$ for all $k > 2$, and $q_k(0) = 0$ for all k, cf. Proposition 6.36.

Now let $\zeta = \eta_1$, the closed 1-form whose integral over $\overline{W^u(p_i^1)}$ is 1 when $i = 1$ and 0 when $i \neq 1$. We have

$$\partial_1^{\mathcal{L}_{\eta_1}}(p_i^1) = 0 \text{ for } i \neq 1$$

$$\partial_1^{\mathcal{L}_{\eta_1}}(p_1^1) = (t^\gamma - t^{\gamma \pm 1}) p^0 \text{ for some } \gamma \in \mathbb{R}$$

$$\partial_2^{\mathcal{L}_{\eta_1}}(p^2) = (t^{\gamma'} - t^{\gamma' \pm 1}) p_2^1 \text{ for some } \gamma' \in \mathbb{R},$$

where $t^\gamma - t^{\gamma \pm 1}$ is invertible in Nov for any $\gamma \in \mathbb{R}$. Therefore,

$$H_k((C_*(f; \mathcal{L}_{\eta_1}), \partial_*^{\mathcal{L}_{\eta_1}})) \approx \begin{cases} 0 & \text{if } k = 0 \\ \text{Nov} \oplus \text{Nov} & \text{if } k = 1 \\ 0 & \text{otherwise,} \end{cases}$$

and we see that $b_1(\eta_1) = 2$, $b_k(\eta_1) = 0$ for all $k \neq 1$, and $q_k(\eta_1) = 0$ for all k.

It's clear that same is true for any closed 1-form cohomologous to any of the basis elements $\{\eta_1, \eta_2, \eta_3, \eta_4\}$ for $H^1_{dR}(S; \mathbb{R})$. To compute the Novikov numbers of a general non exact closed 1-form $\eta \in \Omega^1_{cl}(S, \mathbb{R})$ we will use the Invariance of the Twisted Euler Number (Corollary 4.22), which can be applied to the chain complex $(C_*(f; \mathcal{L}_\zeta), \partial^{\mathcal{L}_\zeta}_*)$ because of Theorems 3.9 and 4.12 and Lemma 4.10.

Assume that $\zeta = a_1\eta_1 + a_2\eta_2 + a_3\eta_3 + a_4\eta_4$ for some $a_1, a_2, a_3, a_4 \in \mathbb{R}$ with $a_j \neq 0$ for some j, i.e. ζ is a non exact closed 1-form, and let $\overline{W(p^2, p_l^1)}$ be the loop on the opposite side of the singular 2-cube from $\overline{W^u(p_j^1)}$ (see Example 6.12). For any $g \in G_{p^2}$,

$$\partial^{\mathcal{L}_\zeta}_2(gp^2) = \sum_{p_l^1 \in Cr_1(f)} \left(t^{\int_0^1 (\gamma^+_{p_l^1 p^2})^*(\zeta)} g - t^{\int_0^1 (\gamma^-_{p_l^1 p^2})^*(\zeta)} g \right) p_l^1,$$

where the coefficient in front of p_l^1 is

$$\left(t^{\int_0^1 (\gamma^+_{p_l^1 p^2})^*(a_1\eta_1+a_2\eta_2+a_3\eta_3+a_4\eta_4)} - t^{\int_0^1 (\gamma^-_{p_l^1 p^2})^*(a_1\eta_1+a_2\eta_2+a_3\eta_3+a_4\eta_4)} \right) g$$

$$= t^A t^B t^C \left(t^{a_j a} - t^{a_j(a\pm 1)} \right) g$$

for some $a, A, B, C \in \mathbb{R}$, because

$$\int_{\gamma^-_{p_l^1 p^2}} \eta_i = \int_{\gamma^+_{p_l^1 p^2}} \eta_i \text{ if } i \neq j \text{ and } \int_{\gamma^-_{p_l^1 p^2}} \eta_j = a \pm 1 \text{ if } \int_{\gamma^+_{p_l^1 p^2}} \eta_j = a.$$

Thus, $\partial^{\mathcal{L}_\zeta}_2(gp^2) \neq 0$ if $g \neq 0$ and $H_2((C_*(f; \mathcal{L}_\zeta), \partial^{\mathcal{L}_\zeta}_*)) \approx 0$. Now let $g \in G_{p_j^1}$. We have $\partial^{\mathcal{L}_\zeta}_1(gp_j^1) =$

$$\left(t^{\int_0^1 (\gamma^+_{p^0 p_j^1})^*(\zeta)} g - t^{\int_0^1 (\gamma^-_{p^0 p_j^1})^*(\zeta)} g \right) p^0$$

$$= \left(t^{\int_0^1 (\gamma^+_{p^0 p_j^1})^*(a_1\eta_1+a_2\eta_2+a_3\eta_3+a_4\eta_4)} - t^{\int_0^1 (\gamma^-_{p^0 p_j^1})^*(a_1\eta_1+a_2\eta_2+a_3\eta_3+a_4\eta_4)} \right) gp^0$$

$$= t^A t^B t^C \left(t^{a_j a} - t^{a_j(a\pm 1)} \right) gp^0$$

6.3 Novikov Homology

for some $a, A, B, C \in \mathbb{R}$, because

$$\int_{\gamma^-_{p^0 p^1_j}} \eta_i = \int_{\gamma^+_{p^0 p^1_j}} \eta_i \text{ if } i \neq j \quad \text{and} \quad \int_{\gamma^-_{p^0 p^1_j}} \eta_j = a \pm 1 \text{ if } \int_{\gamma^+_{p^0 p^1_j}} \eta_j = a.$$

Therefore $\partial_1^{\mathcal{L}_\zeta}$ is surjective when $a_j \neq 0$, and hence $H_0((C_*(f;\mathcal{L}_\zeta), \partial_*^{\mathcal{L}_\zeta})) \approx 0$. Thus,

$$H_k((C_*(f;\mathcal{L}_\zeta), \partial_*^{\mathcal{L}_\zeta})) \approx \begin{cases} 0 & \text{if } k = 0 \\ \text{Nov} \oplus \text{Nov} & \text{if } k = 1 \\ 0 & \text{otherwise,} \end{cases}$$

because $\mathcal{X}_{\mathcal{L}_\zeta}(S) = -2$. This shows that for any non exact closed 1-form ζ we have $b_1(\zeta) = 2$, $b_k(\zeta) = 0$ for all $k \neq 1$, and $q_k(\zeta) = 0$ for all k.

Note that the Novikov inequalities (Theorem 6.37) are nontrivial and distinct from the Morse inequalities in this example. Specifically, the above computation implies that any non exact closed Morse 1-form must have at least two zeros with Morse index 1.

References

1. Abouzaid, M.: Symplectic cohomology and Viterbo's theorem. In: Free Loop Spaces in Geometry and Topology, pp. 271–485 (2015). MR3444367
2. Albers, P., Frauenfelder, U., Oancea, A.: Local systems on the free loop space and finiteness of the Hofer-Zehnder capacity. Math. Ann. **367**(3–4), 1403–1428 (2017). MR3623229
3. Atiyah, M.F., Singer, I.M.: The index of elliptic operators. III. Ann. Math. (2) **87**, 546–604 (1968). MR0236952
4. Audin, M., Damian, M.: Morse theory and Floer homology, Universitext, Springer, London; EDP Sciences, Les Ulis, 2014. Translated from the 2010 French original by Reinie Erné. MR3155456
5. Austin, D.M., Braam, P.J.: Morse-Bott theory and equivariant cohomology. In: The Floer Memorial Volume, pp. 123–183 (1995). MR1362827
6. Banyaga, A.: Sur la structure du groupe des difféomorphismes qui préservent une forme symplectique. Comment. Math. Helv. **53**(2), 174–227 (1978). MR490874 (80c:58005)
7. Banyaga, A.: Some properties of locally conformal symplectic structures. Comment. Math. Helv. **77**(2), 383–398 (2002). MR1915047 (2003d:53134)
8. Banyaga, A., Hurtubise, D.: Lectures on Morse Homology, Kluwer Texts in the Mathematical Sciences, vol. 29. Kluwer Academic Publishers Group, Dordrecht (2004). MR2145196 (2006i:58016)
9. Banyaga, A., Hurtubise, D.E.: Morse-Bott homology. Trans. Amer. Math. Soc. **362**(8), 3997–4043 (2010). MR2608393 (2011e:57054)
10. Banyaga, A., Hurtubise, D.E.: Cascades and perturbed Morse-Bott functions. Algebr. Geom. Topol. **13**(1), 237–275 (2013). MR3031642
11. Barnette, D.: Generating the triangulations of the projective plane. J. Combin. Theory Ser. B **33**(3), 222–230 (1982). MR693361 (84f:57009)
12. Bauer, T., Kitchloo, N., Notbohm, D., Pedersen, E.K.: Finite loop spaces are manifolds. Acta Math. **192**(1), 5–31 (2004). MR2079597
13. Bismut, J.M., Zhang, W.: An extension of a theorem by Cheeger and Müller. With an appendix by François Laudenbach. Astérisque **205**, 7–218 (1992). MR93j:58138
14. Boissonnat, J.-D., Kachanovich, S., Wintraecken, M.: Triangulating submanifolds: an elementary and quantified version of Whitney's method. Discrete Comput. Geom. **66**(1), 386–434 (2021). MR4270647
15. Booss, B., Bleecker, D.D.: Topology and analysis, Universitext. Springer, New York (1985). MR771117
16. Bott, R.: Lectures on Morse theory, old and new. Bull. Amer. Math. Soc. (N.S.) **7**(2), 331–358 (1982). MR84m:58026a

17. Bredon, G.E.: Topology and Geometry, Graduate Texts in Mathematics, vol. 139. Springer, New York (1993). MR1224675 (94d:55001)
18. Burghelea, D., Haller, S.: On the Topology and Analysis of a Closed One Form. I (Novikov's Theory Revisited), Essays on Geometry and Related Topics, vol. 1, 2, pp. 133–175 (2001). MR1929325 (2003h:58030)
19. Burghelea, D., Friedlander, L., Kappeler, T.: On the space of trajectories of a generic gradient like vector field. An. Univ. Vest Timiş. Ser. Mat.-Inform. **48**(1–2), 45–126 (2010). MR2849328 (2012j:58020)
20. Cairns, S.S.: On the triangulation of regular loci. Ann. Math. (2) **35**(3), 579–587 (1934). MR1503181
21. Cairns, S.S.: Polyhedral approximations to regular loci. Ann. Math. (2) **37**(2), 409–415 (1936). MR1503287
22. Cohn, P.M.: Free Rings and their Relations, Second, London Mathematical Society Monographs, vol. 19. Academic Press [Harcourt Brace Jovanovich, Publishers], London (1985). MR800091
23. Cooke, G.E., Finney, R.L.: Homology of Cell Complexes, Based on Lectures by Norman E. Steenrod. Princeton University Press/University of Tokyo Press, Princeton, NJ/Tokyo (1967). MR0219059 (36 #2142)
24. Cornea, O., Ranicki, A.: Rigidity and gluing for Morse and Novikov complexes. J. Eur. Math. Soc. (JEMS) **5**(4), 343.394 (2003). MR2017851
25. de León, M., López, B., Marrero, J.C., Padrón, E.: On the computation of the Lichnerowicz-Jacobi cohomology. J. Geom. Phys. **44**(4), 507–522 (2003). MR1943175 (2003m:53145)
26. Dimca, A.: Sheaves in Topology, Universitext. Springer, Berlin (2004) MR2050072
27. Dold, A.: Lectures on Algebraic Topology, Classics in Mathematics. Springer, Berlin (1995). MR96c:55001
28. Farber, M.S.: Exactness of the Novikov inequalities. Funct. Anal. Its Appl. **19**(1), 40–48 (1985)
29. Farber, M.: Morse-Novikov critical point theory, Cohn localization and Dirichlet units. Commun. Contemp. Math. **1**(4), 467–495 (1999). MR1719699
30. Farber, M.: Topology of Closed One-forms, Mathematical Surveys and Monographs, vol. 108. American Mathematical Society, Providence, RI (2004). MR2034601 (2005c:58023)
31. Farber, M., Ranicki, A.: The Morse-Novikov theory of circle-valued functions and noncommutative localization. Tr. Mat. Inst. Steklova **225**, no. Solitony Geom. Topol. na Perekrest., 381–388 (1999). MR1725953
32. Farber, M., Schütz, D.: Closed 1-forms in topology and dynamics. Uspekhi Mat. Nauk **63**(6(384)), 91–156 (2008). MR2492773
33. Gallais, É.: Combinatorial realization of the Thom-Smale complex via discrete Morse theory. Ann. Sc. Norm. Super. Pisa Cl. Sci. (5) 9(2), 229–252 (2010). MR2731156 (2012a:58021)
34. Goldberg, S.I.: Curvature and Homology. Dover Publications, Mineola, NY (1998). Revised reprint of the 1970 edition. MR1635338
35. Guillemin, V., Sternberg, S.: Geometric Asymptotics. American Mathematical Society, Providence, RI (1977). MR58#24404
36. Hatcher, A.: Algebraic Topology. Cambridge University Press, Cambridge (2002). MR1867354
37. Helffer, B., Sjöstrand, J.: Puits multiples en mécanique semi-classique. IV. Étude du complexe de Witten. Comm. Partial Differential Equations **10**(3), 245–340 (1985). MR780068 (87i:35162)
38. Hilton, P.J., Stammbach, U.: A Course in Homological Algebra, Graduate Texts in Mathematics, vol. 4. Springer, New York (1997). MR1438546
39. Hirsch, M.W.: Differential Topology. Springer, New York (1994). MR96c:57001
40. Hofer, H., Salamon, D.A.: Floer Homology and Novikov Rings, The Floer Memorial Volume, pp. 483–524 (1995). MR1362838
41. Hörmander, L.: The Analysis of Linear Partial Differential Operators. III, Classics in Mathematics. Springer, Berlin (2007). Pseudo-differential operators, Reprint of the 1994 edition. MR2304165

References

42. Hungerford, T.W.: Algebra, Graduate Texts in Mathematics, vol. 73. Springer, New York (1980). Reprint of the 1974 original. MR600654
43. Hurtubise, D.E.: The flow category of the action functional on $\mathcal{L}G_{N,N+K}(\mathbf{C})$. Illinois J. Math. **44**(1), 33–50 (2000). MR1731380 (2001i:57047)
44. Hurtubise, D.E.: Three approaches to Morse-Bott homology. Afr. Diaspora J. Math. **14**(2), 145–177 (2013). MR3093241
45. Jost, J.: Riemannian Geometry and Geometric Analysis, Seventh, Universitext. Springer, Cham (2017). MR3726907
46. Kawakubo, K.: The Theory of Transformation Groups. The Clarendon Press Oxford University Press, New York (1991). MR93g:57044
47. Kronheimer, P., Mrowka, T.: Monopoles and Three-manifolds, New Mathematical Monographs, vol. 10. Cambridge University Press, Cambridge (2007). MR2388043 (2009f:57049)
48. Latour, F.: Existence de 1-formes fermées non singulières dans une classe de cohomologie de de Rham. Inst. Hautes Études Sci. Publ. Math. **80**, 135–194 (1994) (1995). MR1320607 (96f:57030)
49. Latour, F.: A Morse complex on manifolds with boundary. Geom. Dedicata **153**, 47–57 (2011). MR2819662
50. Lee, H.-C.: A kind of even-dimensional differential geometry and its application to exterior calculus. Amer. J. Math. **65**, 433–438 (1943). MR0008495
51. Massey, W.S.: A Basic Course in Algebraic Topology, Graduate Texts in Mathematics, vol. 127. Springer, New York (1991). MR1095046 (92c:55001)
52. McCleary, J.: A User's Guide to Spectral Sequences, Second, Cambridge Studies in Advanced Mathematics, vol. 58. Cambridge University Press, Cambridge (2001). MR1793722 (2002c:55027)
53. Milnor, J.: Morse Theory, Based on Lecture Notes by M. Spivak and R. Wells. Annals of Mathematics Studies, No. 51. Princeton University Press, Princeton, NJ (1963). MR0163331
54. Milnor, J.: Lectures on the h-cobordism Theorem, Notes by L. Siebenmann and J. Sondow. Princeton University Press, Princeton, NJ (1965). MR0190942
55. Mukherjee, A.: Differential Topology, Second. Hindustan Book Agency/Birkhäuser/Springer, New Delhi/Cham (2015). MR3379695
56. Munkres, J.R.: Elementary Differential Topology, Revised, Annals of Mathematics Studies, No. 54. Princeton University Press, Princeton, NJ (1966). Lectures given at Massachusetts Institute of Technology, Fall, 1961. MR198479
57. Nestruev, J.: Smooth Manifolds and Observables, Graduate Texts in Mathematics, vol. 220. Springer, Cham (2020). ©2020. Second edition [of 1930277]. MR4221224
58. Novikov, S.P.: Multivalued functions and functionals. An analogue of the Morse theory. Dokl. Akad. Nauk SSSR **260**(1), 31–35 (1981). MR630459
59. Novikov, S.P.: The Hamiltonian formalism and a multivalued analogue of Morse theory. Uspekhi Mat. Nauk **37**(5(227)), 3–49, 248 (1982). MR676612
60. Ornea, L., Verbitsky, M.: Morse-Novikov cohomology of locally conformally Kähler manifolds. J. Geom. Phys. **59**(3), 295–305 (2009). MR2501742
61. Pajitnov, A.V.: Circle-valued Morse Theory, De Gruyter Studies in Mathematics, vol. 32. Walter de Gruyter & Co., Berlin (2006). MR2319639
62. Palais, R.S.: Local triviality of the restriction map for embeddings. Comment. Math. Helv. **34**, 305–312 (1960). MR123338
63. Pazhitnov, A.V.: On the sharpness of inequalities of Novikov type for manifolds with a free abelian fundamental group. Mat. Sb. **180**(11), 1486–1523, 1584 (1989). MR1034426
64. Pazhitnov, A.V.: On the Novikov complex for rational Morse forms. Ann. Fac. Sci. Toulouse Math. (6) **4**(2), 297–338 (1995). MR1344724
65. Poźniak, M.: Floer homology, Novikov rings and clean intersections. In: Northern California Symplectic Geometry Seminar, pp. 119–181 (1999). MR1736217
66. Qin, L.: On moduli spaces and CW structures arising from Morse theory on Hilbert manifolds. J. Topol. Anal. **2**(4), 469–526 (2010). MR2748215 (2012d:58014)

67. Qin, L.: An application of topological equivalence to Morse theory. J. Fixed Point Theory Appl. **23**(1), Paper No. 10, 38 (2021). MR4198387
68. Ranicki, A.: The algebraic construction of the Novikov complex of a circle-valued Morse function. Math. Ann. **322**(4), 745–785 (2002). MR1905105
69. Ranicki, A.: Circle Valued Morse Theory and Novikov Homology, Topology of High-dimensional Manifolds, No. 1, 2 (Trieste, 2001), pp. 539–569 (2002). MR1937024
70. Reidemeister, K.: Topologie der polyeder und kombinatorische topologie der komplexe, Mathematik und ihre Anwendungen, vol. 17. AkademischeVerlagsgesellschaft, Leipzig (1938)
71. Roman, S.: Advanced Linear Algebra, Third, Graduate Texts in Mathematics, vol. 135. Springer, New York (2008). MR2344656
72. Rotman, J.J.: An Introduction to Homological Algebra, Second, Universitext. Springer, New York (2009). MR2455920
73. Salamon, D.: Morse theory, the Conley index and Floer homology. Bull. London Math. Soc. **22**(2), 113–140 (1990). MR92a:58028
74. Samuel, P.: Algebraic Theory of Numbers, Translated from the French by Allan J. Silberger. Houghton Mifflin Co., Boston, MA (1970). MR0265266
75. Schütz, D.: One-parameter fixed-point theory and gradient flows of closed 1-forms. **K**-Theory **25**(1), 59–97 (2002). MR1899700
76. Schütz, D.: Gradient flows of closed 1-forms and their closed orbits. Forum Math. **14**(4), 509–537 (2002). MR1900172
77. Schütz, D.: Geometric chain homotopy equivalences between Novikov complexes. In: Highdimensional Manifold Topology, pp. 469–498 (2003). MR2048734
78. Schwarz, M.: Morse Homology, Progress in Mathematics, vol. 111. Birkhäuser Verlag, Basel (1993). MR1239174 (95a:58022)
79. Sikorav, J.C.: Points fixes de difféomorphismes symplectiques, intersections de sousvariétés lagrangiennes, et singularités de un-formes fermées, Ph.D. Thesis (1987). Thèse de doctorat dirigée par Laudenbach, François
80. Spanier, E.H.: Algebraic Topology. Springer, New York (1981). Corrected reprint. MR666554 (83i:55001)
81. Spanier, E.: Singular homology and cohomology with local coefficients and duality for manifolds. Pacific J. Math. **160**(1), 165–200 (1993). MR1227511
82. Spivak, M.: A Comprehensive Introduction to Differential Geometry, vol. I. Publish or Perish, Wilmington, Del. (1979). MR82g:53003c
83. Steenrod, N.E.: Homology with local coefficients. Ann. Math. (2) **44**, 610–627 (1943). MR0009114 (5,104f)
84. Steenrod, N.: The Topology of Fibre Bundles, Princeton Mathematical Series, vol. 14. Princeton University Press, Princeton, NJ (1951). MR0039258 (12,522b)
85. Vaisman, I.: On locally conformal almost Kähler manifolds. Israel J. Math. **24**(3–4), 338–351 (1976). MR0418003
86. Vaisman, I.: Remarkable operators and commutation formulas on locally conformal Kähler manifolds. Compos. Math. **40**(3), 287–299 (1980). MR571051 (81j:53063)
87. Vaisman, I.: Locally conformal symplectic manifolds. Int. J. Math. Math. Sci. **8**(3), 521–536 (1985). MR809073
88. Wall, C.T.C.: Differential Topology, Cambridge Studies in Advanced Mathematics, vol. 156. Cambridge University Press, Cambridge (2016). MR3558600
89. Warner, F.W.: Foundations of Differentiable Manifolds and Lie Groups, Graduate Texts in Mathematics, vol. 94. Springer, New York (1983). MR722297 (84k:58001)
90. Weber, J.: The Morse-Witten complex via dynamical systems. Expo. Math. **24**(2), 127–159 (2006). MR2243274
91. Wehrheim, K.: Smooth structures on Morse trajectory spaces, featuring finite ends and associative gluing. Proc. Freedman Fest **18**, 369–450 (2012). MR3084244
92. Weibel, C.A.: An Introduction to Homological Algebra, Cambridge Studies in Advanced Mathematics, vol. 38. Cambridge University Press, Cambridge (1994). MR1269324 (95f:18001)

References

93. Whitehead, G.W.: Elements of Homotopy Theory, Graduate Texts in Mathematics, vol. 61. Springer, New York (1978). MR80b:55001
94. Whitehead, J.H.C.: On C^1-complexes. Ann. Math. (2) **41**, 809–824 (1940). MR2545
95. Whitney, H.: Geometric Integration Theory. Princeton University Press, Princeton, NJ (1957). MR87148
96. Witten, E.: Supersymmetry and Morse theory. J. Differential Geom. **17**(4), 661–692 (1982) (1983). MR84b:58111

Index

K^2, 141, 142
S^1, 4, 6, 15, 18–20, 27, 53, 55, 58, 59, 112, 121, 124, 126, 137–139, 142
T^2, 122, 139, 140
T^n, 121
$W^s(p)$, 1, 14
$W^u(q)$, 1, 14
η-twisted, 3, 6–8, 16, 17, 19, 22, 50, 89, 92, 94, 98, 99, 101, 103–106, 111, 112, 116
η-twisted Morse-Smale-Witten chain complex, 16
η-twisted Morse-Smale-Witten cochain complex, 92, 98
λ_q, 2
$\mathbb{R}P^2$, 4, 20–22, 25, 26, 83, 84, 122, 123
$\mathcal{L}_\xi$, 129, 134, 135
$\mathcal{L}_\zeta$, 8, 9, 129, 134, 135, 137, 142
$\mathcal{X}_\mathcal{L}(X)$, 86
Nov, 128–130, 132, 134–140, 142, 143
Nov(Γ), 128–132, 134, 135
Per$_\xi$, 125–127, 134
Δ^k, 52, 53, 55–57, 63, 79, 81, 132, 133
2-simplex, 64–70

Abouzaid, M., 4
adapted cohomology, 92
algebraic Novikov complex, 125, 132
Audin, M., 61

Banyaga-Hurtubise, 3, 5, 14, 42
Betti number, 9
Bismut, J.M., 103

Bott, R., v
bump function, 72
bundle of abelian groups, 11–13
Burghelea, D., 61, 124, 130

Cartan's formula, 109, 110
cell, v, 53, 59, 62, 63, 71–73, 75, 76, 79, 81
chain homotopy, 42, 44, 45, 125, 131, 133
chain map, 33, 37, 39, 40, 42, 44, 48, 49, 90, 95, 97, 98, 100, 102, 104, 132
circle, 4, 6, 15, 18–20, 27, 53, 58, 59, 112, 121, 137–139, 142
circle valued function, 124–126, 132
Cohn localization, 132
combinatorial Morse theory, 6, 36, 51, 82
compactification, 14, 39, 45, 96
compactified moduli space, 6, 13, 39, 45, 61, 96
conformally equivalent, 93, 94
Conley index theory, 5, 35
constant bundle, 12, 16, 22, 120
covering space, 4, 8, 11, 21, 24, 25, 27, 28, 124, 125, 127, 130–132

Damian, M., 61
deck transformation, 5, 30, 125, 127, 131, 141
de Rham Theorem, 7, 89, 97, 101, 103, 104, 111

Eilenberg, 4, 5, 11, 24, 29, 31, 34, 53, 119, 133
equivariant homology, 24, 34, 53, 133

© The Author(s), under exclusive license to Springer Nature Switzerland AG 2024
A. Banyaga et al., *Twisted Morse Complexes*, Lecture Notes in Mathematics 2361,
https://doi.org/10.1007/978-3-031-71616-4

Euler number, vi, 7, 8, 84, 86, 87, 103, 104, 111, 112, 116, 120, 144
Euler-Poincaré Theorem, 7, 86–88, 104

Farber, M., 124, 125, 131, 132, 137
filtration, 5, 28
Floer homology, v, 3, 5, 36
free loop space, 4
Friedlander, L., 61
fundamental identity, 60, 82

group of periods, 126–128, 130, 131

Haller, S., 124, 130
Hessian, 1
Hodge star operator, 108–110
homotopy unit, 118–120
H-space, vi, 3, 8, 107, 118

index, 1, 2, 14, 17, 25, 104, 124, 130, 136
interior product, 108, 109
Invariance Theorem, 5, 6, 37, 49, 50, 52, 84, 90, 96, 103
isometry, 28, 61, 79–81
isomorphic bundles, 12, 13, 90
Isotopy (Diffeotopy) Extension Theorem, 70, 71

Kappeler, T., 61
Klein bottle, 141
Kronheimer, P., 3

Latour, F., 124, 130
Laudenbach, 61, 96, 103
León, M., 107
Lee form, 7, 94, 112
Lichnerowicz cohomology, vi, 3, 7, 8, 89, 92, 94, 95, 101, 104, 106, 107, 112, 113, 116
local coefficient system, 3–6, 8, 11, 12, 52, 53, 55, 84, 107, 118–123, 128, 129, 133, 134, 136
locally conformal symplectic (LCS), 7, 8, 89, 92–94, 112
López, B., 107

Marrero, J.C., 107
moduli space, 2, 6, 13, 14, 38, 39, 42, 45, 46, 48, 61, 96, 139
Morse Eilenberg Theorem, 31
Morse function, 1
Morse Homology Theorem, 2
Morse-Smale pair, 2
Morse-Smale transversality, 2
Morse-Smale-Witten chain complex, 2
Morse-Smale-Witten cochain complex, 89
Mrowka, T., 3
multivalued function, 124

nondegenerate, 1, 7, 93, 112, 136
Novikov, S.P., 124, 125
Novikov complex, 8, 125, 130–132
Novikov homology, vi, 3, 8, 9, 34, 107, 124–126, 130, 133
Novikov inequalities, 9, 124, 136, 137, 145
Novikov numbers, vi, 3, 9, 107, 135–137, 139, 141, 142, 144
Novikov Principle, 8, 131–133, 137
Novikov ring, 3, 8, 128–132
Novokov theory, 5, 35, 124

orientation, v, 2, 14, 16–18, 20, 25, 28–30, 33, 34, 38–40, 44–48, 57, 58, 82, 83, 89, 90, 92, 97, 101, 115, 116, 139, 142

Padrón, E., 107
Pajitnov, A.V., 29, 124, 130, 131
parallel 1-form, 107, 110
period homomorphism, 125–127, 130, 134
Poincaré Lemma, 7, 89, 99, 105
projective space, 4, 20, 21, 25, 26, 61, 83, 84, 122–124

Ranicki, A., 125, 131
rank (cohomology class), vi, 8, 126–131, 135, 137
regular covering space, 125, 127, 130, 132
regular CW-complex, v, vi, 6, 7, 51–53, 55, 58, 59, 61, 82–84, 86–89, 103
Reidemeister, K., v

Schütz, D., 130, 131
Seiberg-Witten monopole, 3

Index

sheaf cohomology, 7, 89, 104, 106
Sikorav, J.C., 128, 132
simple bundle, 12
simplex, 52, 64–68, 71, 97
singular homology, 52, 53
stable manifold, 1, 2, 14, 29
Steenrod, N.E., v, 5–7, 51–53, 55, 56, 59, 60, 87, 89
Stokes' Theorem, 13, 96, 114
surface of genus two, vi, 9, 113, 122, 142
symplectic cohomology, 4

Thurston, 71
torsion number, 9
torus, 6, 9, 53, 55, 61, 121, 122, 139, 140
triangulation, vi, 6, 29, 35, 36, 51, 61–63, 68, 69, 71–73, 75, 76, 79, 81, 82, 84, 96, 97, 99, 101
twisted Euler number, 7, 8, 86–88, 103, 112, 116, 144
twisted Morse cohomology, 7, 89–92, 94, 96, 98, 99, 101, 104, 106, 107, 111, 112
twisted Morse homology, 3–9, 11, 12, 16, 17, 23, 34

Twisted Morse Homology Theorem, 5, 6, 51, 52, 84, 120, 135, 137
twisted Morse-Smale-Witten chain complex, 16
twisted Morse-Smale-Witten cochain complex, 89
twisted singular boundary operator, 52

universal cover, 5, 21, 25, 27–30, 53, 125, 130, 132, 141
unstable manifold, 1, 5, 6, 14, 16, 30, 34, 36, 38, 44, 51, 58, 59, 61, 63, 84, 89, 92, 96, 97, 99, 101, 103
unstably 0-connected, 99, 101–103
unstably closed, 99, 101–103

Viterbo's Theorem, 4

Witten deformation, 93

Zhang, W., 103

LECTURE NOTES IN MATHEMATICS   Springer

Editors in Chief: J.-M. Morel, B. Teissier;

Editorial Policy

1. Lecture Notes aim to report new developments in all areas of mathematics and their applications – quickly, informally and at a high level. Mathematical texts analysing new developments in modelling and numerical simulation are welcome.

   Manuscripts should be reasonably self-contained and rounded off. Thus they may, and often will, present not only results of the author but also related work by other people. They may be based on specialised lecture courses. Furthermore, the manuscripts should provide sufficient motivation, examples and applications. This clearly distinguishes Lecture Notes from journal articles or technical reports which normally are very concise. Articles intended for a journal but too long to be accepted by most journals, usually do not have this "lecture notes" character. For similar reasons it is unusual for doctoral theses to be accepted for the Lecture Notes series, though habilitation theses may be appropriate.

2. Besides monographs, multi-author manuscripts resulting from SUMMER SCHOOLS or similar INTENSIVE COURSES are welcome, provided their objective was held to present an active mathematical topic to an audience at the beginning or intermediate graduate level (a list of participants should be provided).

   The resulting manuscript should not be just a collection of course notes, but should require advance planning and coordination among the main lecturers. The subject matter should dictate the structure of the book. This structure should be motivated and explained in a scientific introduction, and the notation, references, index and formulation of results should be, if possible, unified by the editors. Each contribution should have an abstract and an introduction referring to the other contributions. In other words, more preparatory work must go into a multi-authored volume than simply assembling a disparate collection of papers, communicated at the event.

3. Manuscripts should be submitted either online at www.editorialmanager.com/lnm to Springer's mathematics editorial in Heidelberg, or electronically to one of the series editors. Authors should be aware that incomplete or insufficiently close-to-final manuscripts almost always result in longer refereeing times and nevertheless unclear referees' recommendations, making further refereeing of a final draft necessary. The strict minimum amount of material that will be considered should include a detailed outline describing the planned contents of each chapter, a bibliography and several sample chapters. Parallel submission of a manuscript to another publisher while under consideration for LNM is not acceptable and can lead to rejection.

4. In general, **monographs** will be sent out to at least 2 external referees for evaluation.

   A final decision to publish can be made only on the basis of the complete manuscript, however a refereeing process leading to a preliminary decision can be based on a pre-final or incomplete manuscript.

   Volume Editors of **multi-author works** are expected to arrange for the refereeing, to the usual scientific standards, of the individual contributions. If the resulting reports can be forwarded to the LNM Editorial Board, this is very helpful. If no reports are forwarded or if other questions remain unclear in respect of homogeneity etc, the series editors may wish to consult external referees for an overall evaluation of the volume.

5. Manuscripts should in general be submitted in English. Final manuscripts should contain at least 100 pages of mathematical text and should always include

   – a table of contents;
   – an informative introduction, with adequate motivation and perhaps some historical remarks: it should be accessible to a reader not intimately familiar with the topic treated;
   – a subject index: as a rule this is genuinely helpful for the reader.
   – For evaluation purposes, manuscripts should be submitted as pdf files.

6. Careful preparation of the manuscripts will help keep production time short besides ensuring satisfactory appearance of the finished book in print and online. After acceptance of the manuscript authors will be asked to prepare the final LaTeX source files (see LaTeX templates online: https://www.springer.com/gb/authors-editors/book-authors-editors/manuscriptpreparation/5636) plus the corresponding pdf- or zipped ps-file. The LaTeX source files are essential for producing the full-text online version of the book, see http://link.springer.com/bookseries/304 for the existing online volumes of LNM). The technical production of a Lecture Notes volume takes approximately 12 weeks. Additional instructions, if necessary, are available on request from lnm@springer.com.

7. Authors receive a total of 30 free copies of their volume and free access to their book on SpringerLink, but no royalties. They are entitled to a discount of 33.3 % on the price of Springer books purchased for their personal use, if ordering directly from Springer.

8. Commitment to publish is made by a *Publishing Agreement*; contributing authors of multiauthor books are requested to sign a *Consent to Publish form*. Springer-Verlag registers the copyright for each volume. Authors are free to reuse material contained in their LNM volumes in later publications: a brief written (or e-mail) request for formal permission is sufficient.

Addresses:
Professor Jean-Michel Morel, CMLA, École Normale Supérieure de Cachan, France
E-mail: moreljeanmichel@gmail.com

Professor Bernard Teissier, Equipe Géométrie et Dynamique,
Institut de Mathématiques de Jussieu – Paris Rive Gauche, Paris, France
E-mail: bernard.teissier@imj-prg.fr

Springer: Ute McCrory, Mathematics, Heidelberg, Germany,
E-mail: lnm@springer.com

SPRINGER NATURE

GPSR Compliance

The European Union's (EU) General Product Safety Regulation (GPSR) is a set of rules that requires consumer products to be safe and our obligations to ensure this.

If you have any concerns about our products, you can contact us on ProductSafety@springernature.com

In case Publisher is established outside the EU, the EU authorized representative is:

Springer Nature Customer Service Center GmbH
Europaplatz 3
69115 Heidelberg, Germany

The manufacturer's authorised representative in the EU is Springer Nature Customer Service Centre GmbH, Europaplatz 3, 69115 Heidelberg, Germany. If you have any concerns regarding our products, please contact ProductSafety@springernature.com

Printed and bound by CPI Group (UK) Ltd, Croydon, CR0 4YY

26/03/2026

02078935-0018